Maths Skills
Builder
Book One

Introduction

Maths Skills Builder Book One is the first in the series of books and assessments papers written by teachers. The exercises and examples have been tried and tested in the classroom and are designed to support parents in preparing their children for state grammar tests, independent school entrance tests and also the Key Stage 2 and 3 new national curriculum framework.

In this series, **Maths Skills Builder Book One** aims to provide essential skills required in securing students' foundations in Maths for all courses at Key Stage 2, 3, state selective tests, 11 + examinations and independent school examinations in Maths. The step-by-step techniques and methods in these workbooks use a clear and development learning process to ensure students' progress.

This book includes;

- How to tackle mental maths
- Written examples maths operations
- Practise examination questions
- Answers

Contents

Place Value

Understanding the value of each number (digit) will dictate how you add, subtract, multiply and divide a number. Without this understanding you will struggle with most of the mathematical concepts. So make sure you spend some extra time familiarising yourselves with this.

Place value before the decimal point:

TH/TH	H/TH	T/TH	TH	H	T	U
2	0	0	0	0	0	0
	3	0	0	0	0	0
		3	4	0	0	0
			3	4	6	0
				5	3	3
					2	3
						3

2000000   (Two million) this number has **7 digits**

300000   (Three hundred thousand) this number has **6 digits**

34000   (Thirty four thousand) this number has **5 digits**

3460   (Three thousand, four hundred and sixty) this number has **4 digits**

533   (Five hundred and thirty three) – this number has **3 digits**

23   (Twenty Three) this number has **2 digits.**

3   (Three) this number has **1 digit**

After the units is a decimal point, then to the right is a mirror image but in fractions. Below is the value of the number 2.31.

U	•	T_{ths}	H_{ths}
2	•	3	1

Two units - 2

Three tenths - $\frac{3}{10}$

One hundredths $\frac{1}{100}$

Ordering Numbers

Example one

Order the following numbers. They are already listed, so they will be easier to order.
Look at the first column.

798
984
428
438

From the first column the highest number is 9, this means that the 984 is the largest number and the one after that is 798. There are two 4's, so we look to the next column to find the third largest number.

428
438

The second column shows that 3 is bigger than the 2 so 438 is the third largest number. The numbers from the largest to the smallest are; 984,798,438,428.

Example Two

Order the following numbers. Using the place value grid, list the numbers according to the value of each digit.

54.2
8.03
56.9
231

H	T	U	•	T$_{ths}$	H$_{ths}$
	5	4	•	2	
		8	•	0	3
	5	6	•	9	
2	3	1	•		

Example Two - continued

Once the numbers are placed in the place value grid, add in 0 where there are empty spaces. This will enable numbers with the same digits to be compared.

Compare column to column. The first column (hundreds) shows that 2 is the largest number, therefore 231 is the larger.

H	T	U	•	Tths	Hths
0	5	4	•	2	0
0	0	8	•	0	3
0	5	6	•	9	0
2	3	1	•	0	0

The second column (tens) is used to find the next largest number. There are two 5's which means the units column is needed to compare the two numbers. The units column shows that 6 is bigger than 4. So 56.9 is the second largest number followed by 54.2.

H	T	U	•	Tths	Hths
0	5	4	•	2	0
0	5	6	•	9	0

This leaves 8.03 which will be the smallest number.

The numbers in order largest to smallest are;

231, 56.9, 54.2, 8.03

Exercise P1 - Write each value in words

a) 7284 _____

b) 284 _____

c) 7535 _____

d) 1487 _____

e) 642 _____

f) 23 _____

g) 715 _____

h) 78 _____

i) 4888 _____

j) 43 _____

Exercise P2 - Provide the standard notation for each value

a) six thousand two hundred and thirteen   _____

b) eight thousand seven hundred and four   _____

c) four thousand seven hundred and forty-two   _____

d) five thousand nine hundred and seventeen   _____

e) eight hundred and three   _____

f) nine thousand and twenty-three   _____

g) nine thousand two hundred and seventy-four   _____

h) two thousand three hundred and twenty-nine   _____

i) eight thousand three hundred and twenty six   _____

j) eight thousand nine hundred and eighty-eight   _____

Exercise P3 - Provide the standard notation for each value

a) 9,000 + 300 + 90 + 5   _____

b) 1,000 + 10   _____

c) 900 + 60 + 6   _____

d) 6,000 + 100 + 40 + 4   _____

e) 3,000 + 600 + 60 + 6   _____

f) 6,000 + 700 + 90 + 4   _____

g) 6,000 + 600 + 20   _____

h) 7,000 +800 + 30 + 8   _____

i) 7,000 + 500 + 70 + 6   _____

j) 9,000 + 200 + 50   _____

Exercise P4 - Determine the place value of the underlined digit

a) 8̲4 = _____

b) 5̲0 = _____

c) 75̲8 = _____

d) 7̲ = _____

e) 514̲8 = _____

f) 799̲1 = _____

g) 8̲93 = _____

h) 6̲0 = _____

i) 4̲10 = _____

J) 22̲7 = _____

k) 6̲410 = _____

l) 20̲27 = _____

Exercise P5 - Determine the place value of the underlined digit

a) 60̲6.9 = _____

b) 6̲.79 = _____

c) 72̲5 = _____

d) 993̲1 = _____

e) 409̲8 = _____

f) 80̲8 = _____

g) 64.6̲ = _____

h) 2̲7.5 = _____

i) 5.8̲3 = _____

j) 3̲.89 = _____

k) 646.7̲ = _____

l) 207̲.5 = _____

Exercise P6 - Round to the underlined digit

a) 5̲322 = _____

b) 676̲9 = _____

c) 9̲810 = _____

d) 6̲295 = _____

e) 4̲849 = _____

f) 5̲138 = _____

g) 260̲9 = _____

h) 8̲747 = _____

i) 6̲311 = _____

j) 6̲726 = _____

k) 173̲11 = _____

l) 6720̲6 = _____

Exercise P7 - Circle the smallest number in each group

a)	b)	c)	d)	e)	f)	g)	h)	i)	j)
673	749	634	984	254	683	303	729	510	290
894	473	692	280	767	556	526	479	888	526
318	312	575	803	451	514	663	116	774	261
258	338	597	461	770	246	582	545	651	336

Exercise P8 - Order the numbers from smallest to largest

a)	b)	c)	d)	e)
513	529	369	854	655
771	182	910	305	273
277	963	665	972	233
770	682	308	370	284

f)	g)	h)	i)	j)
157	228	341	587	453
454	426	428	573	838
507	890	661	625	249
831	183	658	500	565

Exercise P9 - Order the numbers from smallest to largest (with decimal points)

a)
| 45.4 |
| 9.07 |
| 57.9 |
| 134 |

b)
| 488 |
| 49.2 |
| 3.18 |
| 954 |

c)
| 791 |
| 869 |
| 39.9 |
| 654 |

d)
| 6.06 |
| 103 |
| 9.35 |
| 747 |

e)
| 769 |
| 41.0 |
| 355 |
| 422 |

f)
| 807 |
| 3.22 |
| 351 |
| 3.34 |

g)
| 71.2 |
| 445 |
| 59.8 |
| 378 |

h)
| 151 |
| 42.5 |
| 117 |
| 4.49 |

i)
| 64.7 |
| 5.70 |
| 59.2 |
| 31.3 |

j)
| 334 |
| 7.28 |
| 241 |
| 8.38 |

Exercise P10 - Compare the numbers place: > or < or =

> means bigger than
< means smaller than
= means equal to

For example 23 > 5 (23 is bigger than 5)

a) 3907 __ 3   b) 307 __ 6065   c) 8 __ 4   d) 3 __ 4198   e) 77 __ 2

f) 13 __ 481   g) 3 __ 73   h) 5814 __ 4   i) 488 __ 4   j) 15 __ 27

Addition

Keywords that are associated with adding and have the same meaning.

Sum of (sum of 3 and 4 is 7)

Plus (6 plus 8 is 14)

Increase by 10 (add on 10)

Prior Knowledge

Before starting this section, there needs to be an understanding of basic addition facts, like number bonds and how to double and halve numbers. This will help improve understanding of the written methods that are discussed in this book.

2 digits by 1 digit - Basic Addition

Work out 36 plus 7

Using the column method, work out the place value of the numbers. Place the numbers under the specific value as shown below.

T	U
3	6
+	7
4	3
1	

Step 1 - add 6 and 7 first, the answer is 13. Leave the 3 in the units column and carry the 1 to the tens column (see the diagram).

Step 2 - in the tens column add 3 and the 1 that was carried over. The answer is 4.

Exercise A1 - Match the answer with the question by placing the appropriate letter in the given space.

a. 110 + 155 = _____

b. 61 + 14 = _____

c. 103 + 21 = _____

d. 157 + 53 = _____

e. 176 + 104 = _____

f. 199 + 112 = _____

g. 98 + 71 = _____

h. 18 + 93 = _____

i. 70 + 182 = _____

j. 69 + 203 = _____

J = 280

B = 272

I = 169

C = 252

E = 311

D = 265

A = 75

H = 111

G = 124

F = 210

Exercise A2 - What number should be added to the first number to make the second number?

4 + ☐ = 19   10 + ☐ = 16   6 + ☐ = 10   11 + ☐ = 18

12 + ☐ = 12   12 + ☐ = 17   3 + ☐ = 4   7 + ☐ = 10

2 + ☐ = 2   3 + ☐ = 7   15 + ☐ = 18   5 + ☐ = 9

10 + ☐ = 17   6 + ☐ = 12   2 + ☐ = 10   5 + ☐ = 15

3 + ☐ = 6   2 + ☐ = 3   16 + ☐ = 18   3 + ☐ = 5

2 digits by 2 digits

Find the sum of 48 and 86

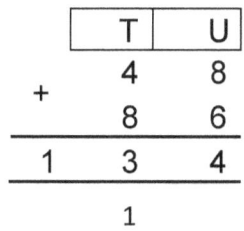

	T	U
	4	8
+	8	6
	1 3	4
	1	

Step 1 - add 8 and 6 first, the answer is 14. Leave the 4 in the units column and carry the 1 to the tens column.

Step 2 - in the tens column add 4, 8 and the 1 that was carried over. The answer is 13.

The 3 will be left in the tens column and the 1 will be placed in the hundreds column.

Exercise A3 - Find the sum

60	72	36	17	13	58	84	43	82
+ 62	+ 74	+ 38	+ 19	+ 15	+ 60	+ 86	+ 45	+ 84

59	52	68	64	89	91	35	80	6
+ 61	+ 54	+ 70	+ 66	+ 91	+ 93	+ 37	+ 82	+ 8

45	10	55	21	87	63	11	27	30
+ 47	+ 12	+ 57	+ 23	+ 89	+ 65	+ 13	+ 29	+ 32

Working out box

Exercise A4 - Find the sum

61	50	26	92	42
34	64	55	38	85
+ 40	+ 14	+ 11	+ 18	+ 30

53	98	95	33	26
98	92	15	27	90
+ 41	+ 64	+ 24	+ 15	+ 76

Working out box

3 digits by 2 digits

Increase 235 by 89

	H	T	U
+	2	3	5
		8	9
	3	2	4
		1	1

Steps 1 - add 5 and 9 first, the answer is 14. Leave the 4 in the units column and carry the 1 to the tens column.

Step 2 - in the tens column add 3, 8 and the 1 that was carried over. The answer is 12. Leave the 2 in the tens column and carry the 1.

Step 3 - in the hundreds column add 2 and the 1 that was carried over. The answer is 3.

Exercise A5 - Find the sum

503	503	143	430	956	431
+ 51	+ 26	+ 46	+ 51	+ 30	+ 30

186	373	333	783	389	772
+ 12	+ 16	+ 52	+ 11	+ 10	+ 16

223	440	981	124	150	433
+ 56	+ 26	+ 12	+ 60	+ 33	+ 13

421	352	875	401	363	108
+ 36	+ 32	+ 10	+ 36	+ 26	+ 30

323	803	189	802	400	246
+ 56	+ 96	+ 10	+ 96	+ 96	+ 30

Working out box

Exercise A6 - Find the sum

236	356	965	360	230
98	62	231	562	398
+ 56	+ 70	+ 51	+ 13	+ 61

538	837	785	86	97
389	939	37	962	662
+ 32	+ 83	+ 816	+ 721	+ 972

Working out box

3 digits by 3 digits

Find the sum of 756 and 689

TH	H	T	U
+	7	5	6
	6	8	9
1	4	4	5
		1	1

Steps 1 - add 6 and 9 first, the answer is 15. Leave the 5 in the units column and carry the 1 to the tens column.

Step 2 - in the tens column add 5, 8 and the 1 that was carried over. The answer is 14. Leave the 4 in the tens column and carry the 1 .

Step 3 - in the hundreds column add 7, 6 and 1 that was carried over. The answer is 14. Place the 1 in the thousands column and the 4 in the hundreds .

Exercise A7 - Find the sum

550 + 332	133 + 643	624 + 244	494 + 402	791 + 200	174 + 524
685 + 201	872 + 125	704 + 153	428 + 420	240 + 643	453 + 223
479 + 510	805 + 111	206 + 363	208 + 611	484 + 201	554 + 145
769 + 220	314 + 231	846 + 121	652 + 224	560 + 118	643 + 114
306 + 542	206 + 172	559 + 310	435 + 312	113 + 441	260 + 417

Working out box

Exercise A8 - Find the sum

```
   272        703        176        984        856
   321        972        930        878        406
+  280     +  682     +  439     +  497     +  715
_____   _____   _____   _____   _____


   121        751        757        186        904
   302        164        868        625        139
+  912     +  386     +  519     +  318     +  807
_____   _____   _____   _____   _____
```

Working out box

4 digits by 4 digits

Add together 5675 and 9378

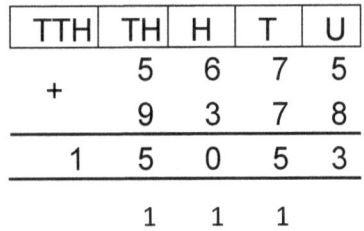

TTH	TH	H	T	U
	5	6	7	5
+				
	9	3	7	8
1	5	0	5	3
	1	1	1	

Steps 1 - add 5 and 8 first, the answer is 13. Leave the 3 in the units column and carry the 1 to the tens column.

Step 2 - in the tens column add 7, 7 and the 1 that was carried over. The answer is 15. Leave the 5 in the tens column and carry the 1.

Step 3 - in the hundreds column add 6,3 and 1 that was carried over. The answer is 10. Leave the 0 in the hundreds column and carry the 1 to the thousands column.

Step 4 - in the ten thousands column add 5, 9 and 1 that was carried over. The answer is 15. Place the 5 in the thousand column and the 1 in the ten thousand column.

Excersise A9 - Find the sum

5450	9313	4285	8801	9803	4011
+ 217	+ 255	+ 601	+ 130	+ 161	+ 853

2260	9107	2296	8715	2390	2870
+ 228	+ 280	+ 301	+ 100	+ 402	+ 114

8441	9469	6632	6674	7122	5276
+ 233	+ 320	+ 267	+ 215	+ 322	+ 611

5167	7432	4004	5897	3119	1340
+ 712	+ 300	+ 500	+ 101	+ 710	+ 153

8593	3510	4246	6804	8105	8679
+ 303	+ 143	+ 640	+ 143	+ 542	+ 210

Working out box

Exercise A10 - Find the sum

```
           986
 4697     1802    8038     3566     5440
 1351     4865    1306     6125      181
+ 4806   + 9290  + 5937   + 7502   + 9035
_____   _____  _____   _____   _____


 3290              1222
 5966      351      955     8002     6250
 1029     1437     7684      252     3498
+ 6638   + 9313   + 7931   + 7262   6239
_____   _____   _____   _____  + 7468
                                    _____


 9970     3169     2853
 1131     2336      225     8765     7087
 9166     7970     3406     9577     2154
+  770   + 2111   + 6347   + 7495   2639
_____   _____   _____   _____  + 2657
                                    _____


 1911     3692              9680
 2352      956     2978      177     5462
 3368     7174     7895     9879     9043
+ 2031   + 1468   + 4037   +  143   + 9862
_____   _____   _____   _____   _____
```

Working out box

18

Exercise A11 - Find the sum

214 + 3115	2724 + 3103	6307 + 2451	137 + 1031	8814 + 1061

6040 + 2417	6571 + 2010	8210 + 1225	120 + 1322	3540 + 1014

1734 + 5115	3950 + 1003	5302 + 176	2504 + 3051	7604 + 1024

563 + 7213	2331 + 1345	6565 + 1112	5854 + 1043	7641 + 158

5542 + 2026	5853 + 103	1165 + 6223	2580 + 4206	4819 + 3010

4360 + 4602	6528 + 310	2784 + 5002	7235 + 1632	3201 + 4067

Working out box

Exercise A12 - Find the sum

5172	9427	6584	3290	5388
2442	7499	2953	1914	7704
8674	9979	3937	2567	4295
+ 7320	6055	+ 6916	+ 8242	7614
	+ 3671			+ 1619

4250	6085	6844	4474	7335
5208	5459	1878	8079	2888
9736	5869	1143	8518	9680
+ 3088	6900	2210	+ 7725	+ 6267
	+ 6060	+ 9907		

1879	6391	5085	7580	1285
8678	8670	8481	7437	6767
9796	6379	4652	2514	1623
4631	+ 8095	8189	+ 5620	6105
+ 7871		+ 4565		+ 3158

6872	6866	1935	2553	8745
2179	9975	7979	7583	6788
2004	6956	4428	9475	2658
8089	+ 5824	2157	8226	7555
+ 1262		+ 3288	+ 9677	+ 2817

Working out box

Exercise A 13 - Using the written methods of addition solve the following and write the answers in the boxes provided.

☐	One thousand and five apples were in the basket. Nine hundred twenty-three are red and the rest are green. How many apples are green?	**Working out box**
☐	Sixty-five peaches are in the basket. Fifty-three more peaches are put in the basket. How many peaches are in the basket now?	
☐	Thirty-five oranges were in the basket. More oranges were added to the basket. Now there are one hundred thirteen oranges. How many oranges were added to the basket?	
☐	Six hundred and thirty-eight pears were in the basket. More pears were added to the basket. Now there are six hundred seventy-seven pears. How many pears were added to the basket?	
☐	One hundred and eighty-six marbles were in the basket. More marbles were added to the basket. Now there are two hundred nine marbles. How many marbles were added to the basket?	
☐	Some oranges were in the basket. Ninety-seven more oranges were added to the basket. Now there are one hundred and forty-four oranges. How many oranges were in the basket before more oranges were added?	
☐	Donald has one hundred and seventy-one marbles and Janet has forty-four marbles. How many marbles do Donald and Janet have together?	
☐	Eight hundred and fifty-six red oranges and fifty-nine green oranges are in the basket. How many oranges are in the basket?	
☐	Some oranges were in the basket. Six hundred and fifteen more oranges were added to the basket. Now there are six hundred ninety-eight oranges. How many oranges were in the basket before more oranges were added?	
☐	David has four hundred and thirty-four more marbles than Brian. Brian has forty-nine marbles. How many marbles does David have?	

Exercise A14 - Find the magic number (same number for the sum diagonally, horizontally and vertically)

10	0	
		6

Magic Number:

	7	11
		4

Magic Number:

		20
	18	16

Magic Number:

						31
	5	7	3	7		32
		1		3	7	26
4				9		31
			8		4	40
3	9		3		4	22
7		2	7	6		25
31	34	28	28	34	21	31

Exercise A16 - Solve

7	+	10	+	10	=	
+		+		+		+
1	+	3	+	2	=	
+		+		+		+
9	+	6	+	5	=	
=		=		=		=
	+		+		=	

1	+	9	+	3	=	
+		+		+		+
4	+	8	+	4	=	
+		+		+		+
8	+	1	+	4	=	
=		=		=		=
	+		+		=	

Exercise - A 17 Solve

Grid 1 — corner: 34

	10	2	9	4	6	6	45
7		7	7		1	10	47
6		7			9		43
	2		3			7	24
	8					8	40
	5				3	9	42
4	2					5	36
45	41	35	37	43	30	46	35

Grid 2 — corner: 24

1		3					35
			2				35
							28
9	5	5		8			39
7	1						40
	1				9		30
2	8		1	8			31
35	28	29	25	49	49	23	39

Solve

Grid 3 — corner: 39

		10					3	45
			1	2	1	5		23
		5			8	7	9	46
						6	1	26
					2	2		41
4	7			3		7		33
9				4			9	49
4		3	8	8			2	36
30	36	34	54	38	31	39	37	28

Grid 4 — corner: 60

1	4							4		51
					3	4		3		48
2		2							9	54
	2	2		5				4		48
			5				3			41
		2				6	1			43
	6	4								59
3		6			1			2		34
	5			2	7	5		2		54
4			5							47
45	46	51	54	49	37	51	40	43	63	51

Adding Decimals

The same concept for addition is used for adding decimals. The decimal point stays in the same place, lined up like the example below.

Remember it is all in the layout, so if the decimal points are not lined up the value of each number will be compromised, which in turn will lead to a wrong answer.

Work out 2.4 + 0.8

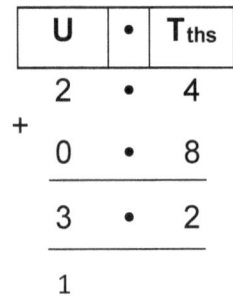

Steps 1 – Line up all the decimal points as shown in the example.

Step 2 - in the tenths column add 4 and 8 the answer is 12. Leave the 2 in the tenths column and carry the 1 to the units column.

Step 3 - in the units column add 2, 0 and 1 that was carried over. The answer is 3. Place the 3 in the units column.

Exercise A20 - Find the sum

8.4 + 3.1	5.2 + 1.2	0.30 + 0.54	7.7 + 4.1	0.16 + 0.26

0.16 + 0.27	1.4 + 1.0	7.4 + 5.4	0.96 + 0.25	8.3 + 4.0

Working out box

Exercise A21 Find the sum

0.079 + 4.955	65 + 2785	0.029 + 1.504	34 + 658	36 + 3005

0.088 + 1.969	0.014 + 4.993	13 + 2955	0.026 + 4.572	76 + 3041

Working out box

Work out 37.43 + 2.014

T	U	•	Tths	Hths	Tths
+ 3	4	•	7	3	
	2	•	0	1	4
		•			

T	U	•	Tths	Hths	Tths
+ 3	4	•	7	3	0
0	2	•	0	1	4
3	6	•	7	4	4

Steps 1 – Line up all the decimal points like the example then add in a zero in the empty spaces.

Step 2 - in the thousandths column add 0 and 4 the answer is 4. Leave the 4 in the thousandths column.

Step 3 - in the hundredths column add 3 and 1. The answer is 4. Place the 4 in the hundredths column.

Step 4 - in the tenths column add 7 and 0. The answer is 7. Place the 7 in the tenths column.

Step 5 - in the units column add 4 and 2. The answer is 6. The 6 is then placed in the units column.

Step 6 - in the tens column add 3 and 0. The answer is 3. Place the 3 in the tenths column.

Exercise A22 - Find the sum

1.7 + 9.5	96 + 3208	0.043 + 0.960
0.051 + 1.763	0.085 + 3.290	7.5 + 101.2
4.2 + 121.6	0.52 + 49.42	3.4 + 193.4

Working out box

Subtraction

Keywords that are associated with subtraction and have the same meaning:-

Difference (difference between 11 and 20 is 9)

Takeaway (9 takeaway 7 is 2)

Decrease by 10 (takeaway 10)
Minus

Prior Knowledge

Before starting this section an understanding of basic subtraction such as number bonds is essential . This will help improve understanding of the written methods that are discussed in this book.

2 digits by 1 digit

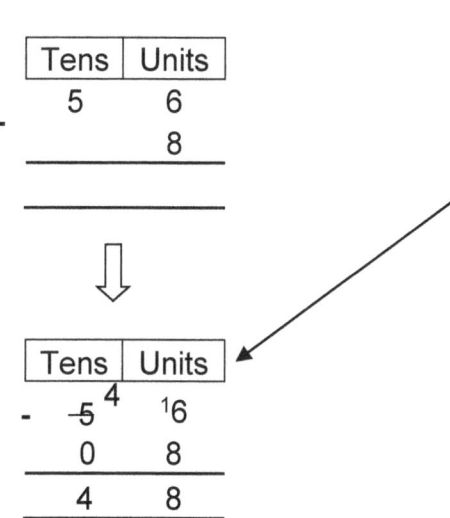

Steps 1 – Start from the units column if the number is smaller (6 is smaller than 8) then borrow 1 ten from the tens column and place it above the 6. This number now reads as 16, as shown in the diagram.

Step 2 – takeaway 8 from 16. The answer is 8. This is then placed in the units column.

Step 3 - in the tens column subtract 0 from 4. The answer is 4. This is placed in the tens column.

2 digits by 2 digit

T	U
4	5
2	9

-

Steps 1 – Start from the units column if the number is smaller (5 is smaller than 9) borrow 1 ten from the tens column and place it above the 5. This number is now 15, as shown in the diagram.

Step 2 – takeaway 9 from 15. The answer is 6, this is then placed in the units column.

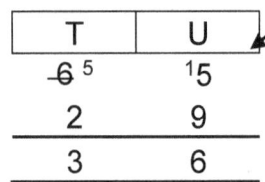

T	U
-6 ⁵	¹5
2	9
3	6

-

Step 3 - in the tens column subtract 2 from 5. The answer is 3. This is placed in the tens column.

Exercise S1 - Find the difference

Working out box

20	40	14	44	50
- 11	- 30	- 7	- 35	- 41

18	17	15	12	21
- 9	- 9	- 7	- 4	- 11

21	15	24	19	14
- 12	- 9	- 14	- 9	- 6

22	12	12	14	22
- 12	- 8	- 7	- 9	- 13

Exercise S2 - Find the difference

57	45	25	76	88	44
- 45	- 15	- 17	- 38	- 73	- 33

22	82	28	28	41	68
- 15	- 63	- 14	- 17	- 11	- 48

67	27	41	79	75	58
- 50	- 11	- 29	- 65	- 32	- 44

44	80	51	27	73	61
- 31	- 65	- 31	- 18	- 65	- 54

82	84	80	39	54	47
- 74	- 14	- 40	- 25	- 23	- 34

Working out box

3 digits by 2 digit

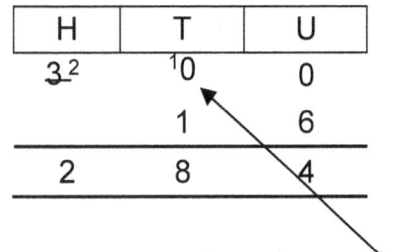

H	T	U
3²	¹0	0
	1	6
2	8	4

Step 1 - Like the previous example, start from the units column. Do the same here but borrow first.

It is not possible to borrow from the tens column because there is a 0 so start from the hundreds column and borrow in stages.

Stage 1: borrow 1 from hundreds and place on top of the 0 in the tens column. This will become the number 10.

Now borrow from the tens column. This becomes 9. The 1 is taken to the units column to make the number 10 .

H	T	U
-3 ²	~~10~~ 9	¹0
	1	6
2	8	4

Step 2 – Now start from the units column takeaway 6 from 10. The answer is 4 this is then placed in the units column.

H	T	U
-3 ²	~~10~~ 9	¹0
0	1	6
2	8	4

Step 3 – In the tens column takeaway 1 from 9. The answer is 8 this is then placed in the tens column.

Step 4 – In the hundreds column takeaway 0 from 2. The answer is 2 this is then placed in the hundreds column.

Exercise S3 - Find the difference

46	207	97	294	720	610
- 36	- 103	- 87	- 106	- 615	- 515

94	294	846	611	210	953
- 84	- 146	- 212	- 110	- 100	- 256

978	200	45	717	922	727
- 805	- 109	- 35	- 400	- 127	- 427

Working out box

Exercise S4 - What number should be added to the first number to make the second number?

361	159	547	461	184
+ ☐	+ ☐	+ ☐	+ ☐	+ ☐
12	90	114	105	97

Working out box

552	295	871	709	559
+ ☐	+ ☐	+ ☐	+ ☐	+ ☐
58	58	31	38	116

603	594	520	211	238
+ ☐	+ ☐	+ ☐	+ ☐	+ ☐
78	22	112	46	83

804	184	466	344	259
+ ☐	+ ☐	+ ☐	+ ☐	+ ☐
57	100	27	114	98

Exercise S5 - Match the answer with the question

a. 155 - 131 = _____ •
b. 136 - 25 = _____ •
c. 84 - 46 = _____ •
d. 27 - 21 = _____ •
e. 127 - 78 = _____ •
f. 186 - 56 = _____ •
g. 153 - 87 = _____ •
h. 183 - 13 = _____ •
i. 38 - 20 = _____ •
j. 87 - 87 = _____ •

• A = 18
• G = 130
• I = 0
• C = 66
• B = 6
• E = 111
• H = 38
• D = 49
• J = 24
• F = 170

Exercise S6 - Find the difference

				Working out box

```
   1108        2804        5504        6803
 -  193      -  103      -  511      -  632
 _____      _____      _____      _____


   8209        3401        9505        7706
 -  147      -  251      -  400      -  441
 _____      _____      _____      _____


   2705        4907        3302        9101
 -  193      -  582      -  341      -  161
 _____      _____      _____      _____


   3402         303        5508        1401
 -  252      -  123      -  175      -  490
 _____      _____      _____      _____


   4104        1800        3901        5004
 -  134      -  410      -  591      -  221
 _____      _____      _____      _____
```

Exercise S7 - Find the difference

34000 - 926	56000 - 159	63000 - 340	5000 - 464	40000 - 280

27000 - 210	39000 - 433	2000 - 242	17000 - 876	8000 - 801

17000 - 142	6000 - 118	8000 - 736	65000 - 530	5000 - 776

48000 - 958	30000 - 230	87000 - 401	3000 - 243	98000 - 488

Working out box

Exercise S8 - Solve

Eight hundred and three balls are in the basket. Thirty are red and the rest are green. How many balls are green?

Seven hundred and ninety marbles are in the basket. Seventy-nine are red and the rest are green. How many marbles are green?

Janet has twenty-eight plums. Audrey has seven hundred and fifty-four plums. How many more plums does Audrey have than Janet?

Some balls were in the basket. Thirty-four balls were taken from the basket. Now there are two hundred and twenty balls. How many balls were in the basket before some of the balls were taken?

Five hundred and fifty-one balls were in the basket. Some of the balls were removed from the basket. Now there are five hundred twenty-six balls. How many balls were removed from the basket?

Five hundred and sixteen oranges were in the basket. Some of the oranges were removed from the basket. Now there are four hundred and thirty-eight oranges. How many oranges were removed from the basket?

Eight hundred and seventy-eight peaches are in the basket. Sixty-five are red and the rest are green. How many peaches are green?

Marin has three hundred and fifty-four fewer peaches than Brian. Brian has four hundred and eighteen peaches. How many peaches does Marin have?

Five hundred and eighty-eight apples are in the basket. Eighty-four apples are taken out of the basket. How many apples are in the basket now?

Three hundred and thirty-six marbles are in the basket. Forty-nine are red and the rest are green. How many marbles are green?

Subtracting Decimals

The concept for adding decimals is also applied for subtracting them, the decimal point stays in the same place. Line the decimal point up as shown in the example below. It is all in the layout.

If the decimal point is not lined up the value of each number will be compromised, which means the answer will be wrong.

Step 1 – Line up all the decimal points, as shown in the example.

Step 2 - in the tenths column takeaway 4 from 9. The answer is 5. Leave the 5 in the tenths column.

Step 3 - in the units column takeaway 1 from 2. The answer is 1. the 1 is placed in the units column.

U	•	Tths
2	•	9
- 1	•	4
3	•	5

Work out 37.43 - 2.014

Step 1 – Line up all the decimal points like in the example then add in zeros in the empty spaces.

T	U	•	T_{ths}	H_{ths}	T_{ths}
3	4	•	7	3	
-	2	•	0	1	4
		•			

Step 2 - in the thousandths column there is a 0, this is smaller than 4, which means that 1 hundreth will be borrowed from the hundredths column and placed above the 0. This number now is 10 like in the diagram. Takeaway 4 from 10 the answer is 6 this is then placed in the tths column.

T	U	•	T_{ths}	H_{ths}	T_{ths}
3	4	•	7	²3̶	¹0
- 0	2	•	0	1	4
3	2	•	7	1	6

Step 3 - in the hundredths column subtract 1 from 2 the answer is 1. Place the 1 in the hundredths column.

Step 4 - in the tenths column subtract 0 from 7 the answer is 7. Place the 7 in the tenths column.

Step 5 - in the units column subtract 2 from 4 the answer is 2. Place the 2 in the units column.

Step 6 - in the tens column subtract 0 from 3 the answer is 3. Place the 3 in the tenths column.

Exercise S9 - Find the difference

				Working out box
7.4 - 4.3	0.29 - 0.18	0.081 - 0.018	0.080 - 0.017	
3.0 - 1.2	0.48 - 0.45	9.7 - 7.9	0.082 - 0.078	
49 - 27	0.081 - 0.072	0.037 - 0.025	0.97 - 0.85	
0.080 - 0.061	0.056 - 0.044	9.8 - 4.7	0.79 - 0.66	
7.1 - 5.7	0.044 - 0.039	29 - 14	3.5 - 1.3	
0.68 - 0.63	70 - 45	9.3 - 4.2	9.6 - 7.12	
88 - 79	6.3 - 3.6	80 - 45	0.046 - 0.011	

Exercise S10 - Find the difference

69 - 17 = _____   69 - 0.013 = _____

8.1 - 0.075 = _____   0.092 - 0.034 = _____

91 - 3.4 = _____   0.62 - 0.46 = _____

0.065 - 0.045 = _____   0.68 - 0.26 = _____

0.037 - 0.016 = _____   73 - 33 = _____

0.068 - 0.063 = _____   0.24 - 0.013 = _____

64 - 1.8 = _____   0.82 - 0.69 = _____

0.036 - 0.033 = _____   98 - 4.5 = _____

0.037 - 0.025 = _____   85 - 11 = _____

83 - 2.5 = _____   0.096 - 0.063 = _____

Working out box

Exercise S11 - Solve

47	-	14	-	19	=	
-		-		-		-
15	-	1	-	8	=	
-		-		-		-
20	-	5	-	8	=	
=		=		=		=
	-		-		=	

46	-	15	-	17	=	
-		-		-		-
11	-	4	-	6	=	
-		-		-		-
17	-	8	-	2	=	
=		=		=		=
	-		-		=	

3	-	2	+	5	=	
-		+		-		+
2	+	1	-	1	=	
+		-		+		+
4	-	1	+	4	=	
=		=		=		=
	+		+		=	

9	-	8	+	3	=	
-		+		-		+
8	+	6	-	1	=	
+		-		+		+
9	-	1	+	10	=	
=		=		=		=
	+		+		=	

Find the secret trail

5	7	1	7
2	2	8	10
(44)	1	1	3
5	8	9	4

(10)

(29)	6	4	5
9	3	5	9
8	1	5	8
7	5	2	5

(1)

3	3	6	9
(41)	9	2	8
6	6	3	3
6	1	4	1

(3)

Multiplication

The words associated with multiplying:

Product - product of 3 and 2 (3 x 2) is 6

Times – 6 times 3 is (6x3) is 18 Double

means multiply by 2

Lots of

Times table

2 digits by 1 digit

Method 1

```
H   T   U
    4   5
x
        7
        5
      3
```

Step 1 – start with the units columnand multiply 7 by 5. The answer is 35. Leave the 5 in the units column and place the 3 under the tens column, as shown in the diagram.

Step 2 – multiply 7 by 4. The answer is 28. Now add the 3 below the tens column. The answer is 31. Place the 3 in the hundreds column and the 1 in the tens.   The answer is 315.

```
H   T   U
    4   5
x
        7
3   1   5

    3
```

2 digits by 1 digit - Method 2 – Grid Method

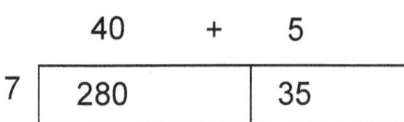

	40	+	5
7	280		35

H	T	U
	2	8	0
+		3	5
	3	1	5

1

Step 1 – multiply 7 by 4. The answer is 28. Now add on a 0 (because there is 1 zero in the calculation 7x 40). The answer is 280.

Step 2 – work out 7 x 5 = 35

Step 3 – add together 280 and 35 using the column method. The answer is 315.

Exercise M1 - Find the product

Working out box

8 × 7 = _____   10 × 5 = _____   13 × 5 = _____

12 × 5 = _____   10 × 7 = _____   12 × 3 = _____

10 × 9 = _____   14 × 8 = _____   8 × 6 = _____

11 × 2 = _____   8 × 5 = _____   11 × 3 = _____

12 × 6 = _____   9 × 7 = _____   15 × 4 = _____

14 × 4 = _____   10 × 2 = _____   11 × 8 = _____

3 × 8 = _____   11 × 7 = _____   5 × 8 = _____

13 × 6 = _____   8 × 2 = _____   13 × 8 = _____

Exercise M2 - Complete the table

x	8	6	11	10	2
4					
15					
3					
5					
10					

x	7	9	15	8	14
9					
12					
10					
6					
8					

x	9	6	8	14	5
2					
11					
12					
8					
13					

x	9	6	12	7	5
6					
9					
4					
12					
3					

Working out box

42

Exercise M3 - Find the product

38 × 2	66 × 4	22 × 8	23 × 4	69 × 5

4 × 4	78 × 8	14 × 4	69 × 3	62 × 5

87 × 2	9 × 5	89 × 8	89 × 5	12 × 7

59 × 3	16 × 9	34 × 9	6 × 8	88 × 8

Working out box

Exercise M4 - Find the product

```
   784          690          649          609          168
×    5        ×    7       ×    8       ×    4       ×    8
_____       _____       _____       _____       _____

_____       _____       _____       _____       _____
```

```
   303          681          906          473          846
×    9        ×    7       ×    8       ×    5       ×    6
_____       _____       _____       _____       _____

_____       _____       _____       _____       _____
```

```
   351          708          869          811          947
×    7        ×    9       ×    5       ×    7       ×    2
_____       _____       _____       _____       _____

_____       _____       _____       _____       _____
```

```
   211          671          547          338          648
×    5        ×    7       ×    6       ×    6       ×    6
_____       _____       _____       _____       _____

_____       _____       _____       _____       _____
```

Working out box

Exercise M5 - Find the product

3200	1322	1010	2201	2332
× 3	× 2	× 4	× 6	× 3

2212	1340	1121	1100	2404
× 4	× 8	× 4	× 4	× 5

4141	1120	1256	2120	1123
× 7	× 4	× 4	× 4	× 6

3301	4344	1313	1110	6809
× 3	× 8	× 3	× 5	× 9

Working out box

2 digits by 2 digits   Calculate the product of 96 and 48

Method 1

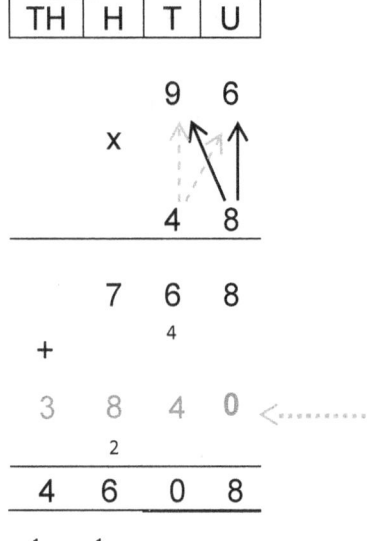

TH	H	T	U
		9	6
	x		
		4	8
	7	6	8
		4	
+			
3	8	4	0
	2		
4	6	0	8
1	1		

Step 1 – start with units column, multiply 8 by 6. The answer is 48. Leave the 8 in the units column and place the 4 under the tens column.

⇩

Step 2 – now multiply 8 by 9. The answer is 72. Add the 4 below the tens column. The answer is 76. Place the 7 in the hundreds column and the 6 in the tens.

⇩

Step 3 – before moving onto the 4 add a zero in the answer section to indicate the tens column.

Starting with the tens column, multiply 4 by 6. The answer is 24. Leave the 4 in the tens column next to the 0 and place the 2 under the hundreds column.

Multiply 4 by 9. The answer is 36. Now add the 2 under the hundreds column. The answer is 38, so place the 8 in the hundreds column and 3 in the thousands column.

⇩

Step 4 – Using the addition method add 768 and 3840.

1) Add 8 and 0, the answer is 8. Place the 8 in the units column.

2) Add 6 and 4. The answer is 10. Place the 0 in the tens column. Now carry the 1 and place under the hundreds column.

3) Add 7 and 8 and the 1 under the hundreds column. The answer is 16. Place the 6 in the hundreds column and carry the 1 under the thousands column.

4) Add 3 and the 1 under the thousands column. The answer is 4. Place the 4 in the thousand column.

Method 2 - the Grid Method

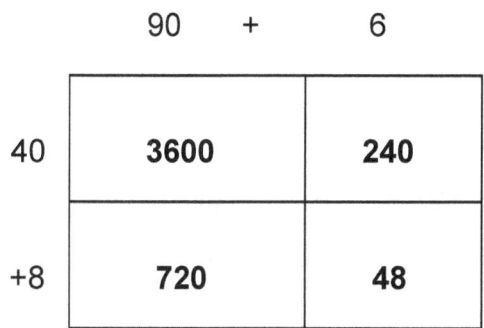

	90 +	6
40	3600	240
+8	720	48

Step 1 – multiply 9 by 4. The answer is 36. Add on 2 0s (because there are 2 zeros in the calculation 90 x 40). The answer is 3600.

Step 2 – multiply 6 by 40. Multiply 6 by 4, the answer is 24. Now add on a zero (because there is a zero in the calculation 6 x 40). The answer is **240.**

Step 3 - multiply 8 by 90. Multiply 8 by 9, the answer is 72. Now add on a zero (because there is a zero in the calculation 8 x 90). The answer is 720.

Step 4 - multiply 8 by 6 so the answer is 48.

Step 5 - add together 3600, 240, 720 and 48 using the column method. Tthe answer is 4608.

Exercise M6 - Find the product.

```
    11          35           9           92          72
  ×  11       ×  11       ×  40        ×  83       ×  80
  _____      _____      _____       _____      _____


    25          43          31           35          49
  ×  89       ×  31       ×  76        ×  53       ×  32
  _____      _____      _____       _____      _____


    20          39          47            1          48
  ×  30       ×  67       ×  80        ×  10       ×  37
  _____      _____      _____       _____      _____


    16          63          44           10          85
  ×  67       ×  73       ×  69        ×  13       ×  60
  _____      _____      _____       _____      _____
```

Working out box

Exercise M7 - Find the product

41	61	97	15	10
× 42	× 61	× 96	× 87	× 89

43	28	82	65	39
× 50	× 61	× 78	× 24	× 92

64	416	661	843	848
× 51	× 85	× 17	× 19	× 19

405	166	174	143	566
× 63	× 21	× 69	× 83	× 16

Working out box

Exercise M8 - Find the product

760	412	348	314	699
× 220	× 249	× 327	× 762	× 416

226	677	324	752	247
× 793	× 552	× 28	× 150	× 432

3010	3204	2314	2222	2111
× 23	× 52	× 82	× 36	× 47

1010	2225	2001	2112	2012
× 85	× 74	× 94	× 99	× 59

Working out box

Exercise M9 - Find the product

```
    472        226
  ×  795     ×  724
  _____    _____
```

```
    326        271
  ×  760     ×  394
  _____    _____
```

```
    502        520
  ×  404     ×  651
  _____    _____
```

```
    883        745
  ×  104     ×  300
  _____    _____
```

```
    177        162
  ×  855     ×  687
  _____    _____
```

Working out box

Multiplying Decimals

Multiplying numbers with decimal points

Multiply 2.35 by 4.6

To turn the numbers into whole numbers, take out the decimal points first. Then multiply and finally adjust for the decimal point.

Method 1 - multiplying 235 by 46

Follow the steps in the multiplying section. Use either long multiplication or the grid method.

H	T	U
2	3	5

				4	6
	1	4	1	0	
		2	3		
	9	4	0	0	
	1	2			
1	0	8	1	0	

Adjust the number by 3 decimal places (2.35 has 2 decimal places and 4.6 has 1 decimal place) so the answer is 10.81.

Exercise M10 - Find the product

28 × 0.007 = _____   74 × 0.6 = _____

47 × 0.6 = _____   28 × 0.5 = _____

20 × 0.002 = _____   39 × 0.006 = _____

35 × 0.4 = _____   88 × 0.004 = _____

43 × 0.2 = _____   52 × 0.06 = _____

96 × 0.8 = _____   13 × 0.8 = _____

46 × 0.2 = _____   45 × 0.07 = _____

65 × 0.09 = _____   20 × 0.001 = _____

88 × 0.09 = _____   98 × 0.008 = _____

72 × 0.001 = _____   74 × 0.002 = _____

28 × 0.6 = _____   88 × 0.003 = _____

44 × 0.8 = _____   30 × 0.004 = _____

54 × 0.02 = _____

Working out box

Multiplying by ten

The number moves to the left by 1 place, which gives the illusion that the decimal place has moved to the right.

For example 3 x 10

H	T	U	•	T$_{ths}$	H$_{ths}$
		3	•	0	

If we are multiplying by 10, the number moves to the left by 1. In the empty space add a 0. This gives the illusion of the decimal point moving to the right.

This method can be used in multiplying numbers by 10, 100, 1000....

Mathematically it should be understood that the number actually moves.

Moving the decimal point is used as a quick and easy method.

H	T	U	•	T$_{ths}$	H
	3	0	•	0	

A quick way to multiply by 10, 100, and 1000 is to do the following.

To multiply by 10, add a 0   **3 x 10 = 30**

To multiply by 100, add on 00   **5 x 100 = 500**

To multiply by 1000 we add on 000   **6 x 1000 = 6000**

To multiply by 10000, add on 0000   **9 x 10000 =90000**

Using this rule we can multiply numbers with zeros (like the grid method above).

30 x 2100

First calculate 3 x 21 = 63

Now count how many zeros are in the calculation (3 zeros) and add them on the end of 63.

So 30 x 2100 = 63000

For example

H	T	U	•	Tths	Hths
		6	•	9	7

6.97 x 10

If we multiply by 10, the number moves to the left by 1.

H	T	U	•	Tths	Hths
	6	9	•	7	

This gives an illusion of the decimal point moving to the right by 1.
This method can be used in calculating numbers being multiplied by 10, 100, 1000......

To multiply by 10, move the decimal point to the right by 1   3.5 x 10 = 3.

5 (decimal point moves 1 place) answer is 35.

To multiply by 100 move the decimal point to the right by 2   5.3 x 100 = 5 . 3   0

(decimal point moves 2 spaces) the answer is 530

Multiplying by 1000 – move by 3 decimal places

Multiplying by 10000 - move by 4 decimal places

Exercise M11 - Multiply

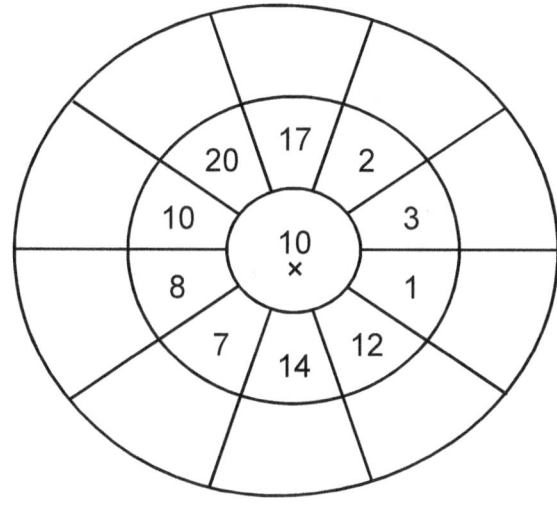
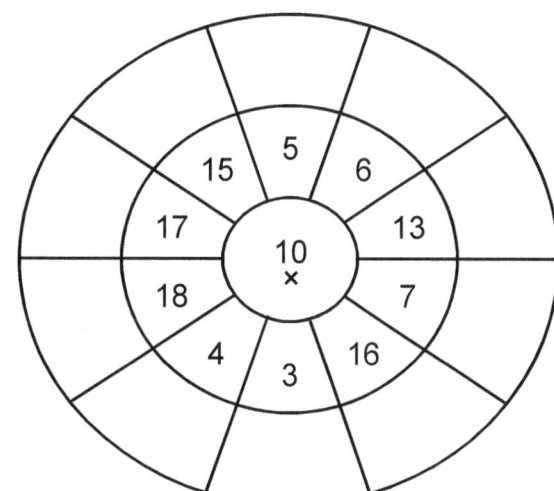
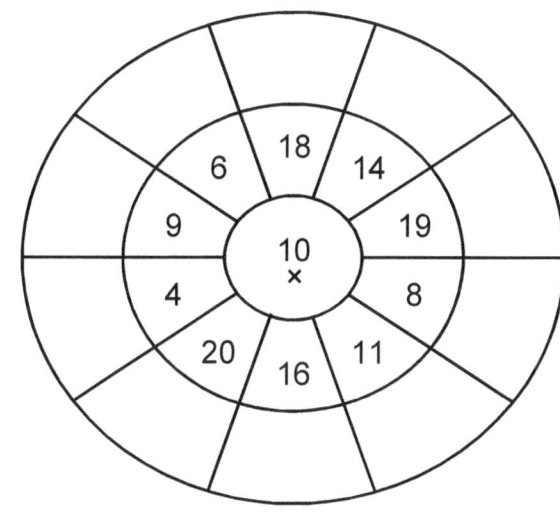
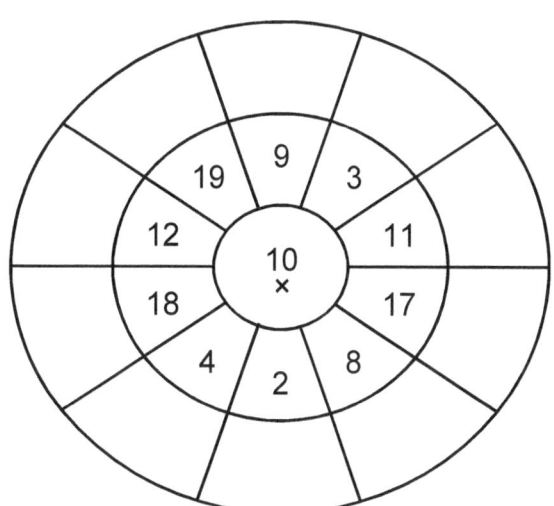

Working out box

Exercise M12- Find the product.

53 × 40 = _____   70 × 20 = _____

37 × 60 = _____   14 × 20 = _____

93 × 80 = _____   59 × 20 = _____

62 × 40 = _____   35 × 80 = _____

60 × 30 = _____   66 × 30 = _____

79 × 10 = _____   71 × 70 = _____

48 × 40 = _____   29 × 60 = _____

94 × 40 = _____   45 × 90 = _____

61 × 60 = _____   65 × 40 = _____

32 × 60 = _____   28 × 50 = _____

Working out box

Exercise M13 - Find the product

27 × 300 = _____   61 × 500 = _____

90 × 500 = _____   72 × 300 = _____

18 × 500 = _____   78 × 300 = _____

67 × 200 = _____   59 × 500 = _____

93 × 500 = _____   51 × 900 = _____

12 × 600 = _____   80 × 700 = _____

99 × 200 = _____   92 × 300 = _____

88 × 500 = _____   12 × 300 = _____

65 × 900 = _____   47 × 900 = _____

83 × 100 = _____   96 × 400 = _____

Working out box

Exercise M14 - Find the product

18 × 2000 = _____

73 × 2000 = _____

42 × 8000 = _____

84 × 8000 = _____

22 × 6000 = _____

33 × 4000 = _____

60 × 2000 = _____

71 × 9000 = _____

69 × 2000 = _____

18 × 3000 = _____

94 × 9000 = _____

55 × 2000 = _____

58 × 9000 = _____

20 × 1000 = _____

29 × 7000 = _____

93 × 9000 = _____

32 × 9000 = _____

71 × 3000 = _____

21 × 4000 = _____

46 × 6000 = _____

Working out box

If we divide by ten, the number moves to the right by 1 place. This gives the illusion that the decimal place moves to the left.

For example 3 ÷ 10

H	T	U	•	Tths	Hths
		3	•	0	

Dividing by 10 the decimal point moves to the left by 1.
This method can be used for numbers being divided by 10, 100, 1000 ....

H	T	U	•	Tths	Hths
		0	•	3	

To divide by 10, move the decimal point to the left by 1   **3.5 ÷ 10 = 3 . 5**

The answer is 0.35

To divide by 100 move the decimal point to the left by 2   **5.3 ÷ 100 = The 5 . 3**

answer is 0.053

Dividing by 1000 – move 3 decimal places to the left

Dividing by 10000 - move 4 decimal places to the left

Exercise D1 - Divide by 10

333 ÷ 10 = ____   175 ÷ 10 = ____

529 ÷ 10 = ____   676 ÷ 10 = ____

807 ÷ 10 = ____   202 ÷ 10 = ____

892 ÷ 10 = ____   297 ÷ 10 = ____

499 ÷ 10 = ____   205 ÷ 10 = ____

Working out box

Exercise D2 - Divide by 100

808 ÷ 100 = ____   470 ÷ 100 = ____

720 ÷ 100 = ____   176 ÷ 100 = ____

953 ÷ 100 = ____   770 ÷ 100 = ____

463 ÷ 100 = ____   652 ÷ 100 = ____

441 ÷ 100 = ____   622 ÷ 100 = ____

Working out box

Exercise D3 - Divide by 1000

927 ÷ 1000 = _____   727 ÷ 1000 = _____

732 ÷ 1000 = _____   101 ÷ 1000 = _____

206 ÷ 1000 = _____   496 ÷ 1000 = _____

960 ÷ 1000 = _____   554 ÷ 1000 = _____

513 ÷ 1000 = _____   259 ÷ 1000 = _____

Working out box

Dividing

Words which also mean divide are, share into, quotient and how many times does it go into.

Exercise D4 - To find the quotient, use your multiplication facts:-
For example $3 \times 2 = 6$ so $6 \div 2 = 3$ and $6 \div 3 = 2$

$28 \div 7 = $ ___	$20 \div 5 = $ ___	$12 \div 6 = $ ___	$48 \div 6 = $ ___
$72 \div 8 = $ ___	$56 \div 8 = $ ___	$12 \div 3 = $ ___	$24 \div 3 = $ ___
$40 \div 8 = $ ___	$25 \div 5 = $ ___	$18 \div 6 = $ ___	$6 \div 2 = $ ___
$49 \div 7 = $ ___	$9 \div 3 = $ ___	$24 \div 4 = $ ___	$15 \div 5 = $ ___
$35 \div 5 = $ ___	$64 \div 8 = $ ___	$36 \div 6 = $ ___	$56 \div 7 = $ ___
$9 \div 9 = $ ___	$16 \div 8 = $ ___	$24 \div 8 = $ ___	$21 \div 3 = $ ___
$45 \div 9 = $ ___	$30 \div 5 = $ ___	$4 \div 4 = $ ___	$6 \div 6 = $ ___
$42 \div 6 = $ ___	$14 \div 7 = $ ___	$63 \div 7 = $ ___	$30 \div 6 = $ ___
$32 \div 8 = $ ___	$7 \div 1 = $ ___	$12 \div 4 = $ ___	$42 \div 7 = $ ___
$8 \div 4 = $ ___	$16 \div 2 = $ ___	$15 \div 3 = $ ___	$18 \div 2 = $ ___

Working out box

Exercise D5 - Complete the table

÷	61	79	31	22	86
3					
7					
5					
4					
9					

÷	84	98	80	45	43
6					
2					
1					
4					
7					

Working out box

Short division - Dividing into all place values exactly

Using multiplication facts you should be able to divide.

Example

$4 \times 8 = 32$

Two division

facts $32 \div 8 = 4$

$32 \div 4 = 8$

Use what you know to work out what you do not know. These facts will help you divide. You can use these facts for long division.

Share 3 into 78:

The 3 is the divisor which is placed outside and the 78 which is the

dividend inside.

Write as follows

```
        2      6
  3 |   7     ¹8
```

Step 1 – how many 3's go into 7 ? It is 2 with a remainder of 1.

The 2 is placed directly above the 7 and the remainder is placed directly above the 8 to make the number 18.
Step 2 – How many 3's go into 18 ? The answer is 6. Place the 6 directly on top of the 18.

The answer to 78 ÷3 = 26

Also 3 x 26 = 78

Exercise D6 - Find the quotient

$5\overline{)90}$ $3\overline{)78}$ $7\overline{)280}$ $8\overline{)72}$ $5\overline{)760}$

$5\overline{)460}$ $7\overline{)938}$ $8\overline{)208}$ $9\overline{)513}$ $7\overline{)154}$

$2\overline{)532}$ $9\overline{)18}$ $3\overline{)33}$ $1\overline{)76}$ $2\overline{)282}$

$5\overline{)30}$ $6\overline{)84}$ $6\overline{)78}$ $2\overline{)70}$ $5\overline{)980}$

$4\overline{)72}$ $8\overline{)64}$ $5\overline{)50}$ $4\overline{)44}$ $6\overline{)966}$

Working out box

Exercise D7 - Find the quotient

$7\overline{)196}$ $7\overline{)266}$ $8\overline{)200}$ $3\overline{)183}$ $6\overline{)282}$

$5\overline{)235}$ $6\overline{)384}$ $2\overline{)58}$ $4\overline{)176}$ $8\overline{)440}$

$3\overline{)273}$ $5\overline{)190}$ $7\overline{)616}$ $7\overline{)238}$ $6\overline{)96}$

$7\overline{)350}$ $5\overline{)255}$ $2\overline{)118}$ $7\overline{)133}$ $1\overline{)54}$

$7\overline{)217}$ $1\overline{)87}$ $8\overline{)728}$ $8\overline{)288}$ $5\overline{)165}$

Working out box

Short division – dividing by a single digit and finding remainders

Method 1   **8496 ÷ 6**

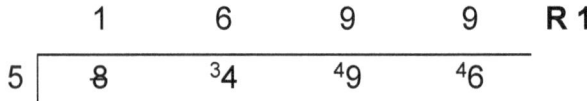

```
        1       6       9       9     R 1
   ┌────────────────────────────────────
 5 │    8     ³4      ⁴9      ⁴6
```

Step 1 – How many 5's go into 8? It is 1 with a remainder of 3. The 1 is placed directly above the 8 and the remainder of 3 is placed on top of the 4 to make the number 34.

Step 2 – How many 5's go into 34? The answer is 6 with a remainder of 4. Place the 6 directly on top of the 34. The remainder of 4 is placed on top of the 9 to make the number 49.

Step 3 - How many 5's go into 49? The answer is 9 with a remainder of 4. Place the 9 on top of the 49 and the remainder on top of the 6 to make the number 46.

Step 4 - How many 5's go into 46? The answer is 9 with a remainder of 1. Place the 9 directly on top of 46 and place the remainder directly next to the 9, as shown in the diagram.

Method 2 using multiplication and subtraction

(This method can be used for divisors greater than 1)

8496 ÷ 6

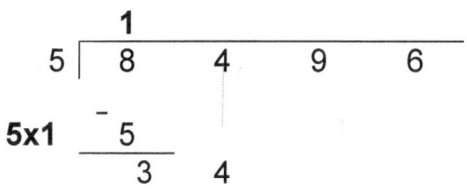

```
        1
   5 | 8    4    9    6
  -
5x1   5
      3    4
```

Step 1 – How many 5's go into 8? It is 1 whole which is placed at the top of 8, at this point any remainders are ignored.

Step 2 – The 1 is then multiplied by the 5 (divisor) (5 x 1). The answer 5 is then placed under the 8. Then underline as shwon in the diagram.

Now subtract 5 from 8. The answer is 3 which directly placed under the 5 (as shown in the diagram). Now bring down the next number 4 to make the number 34 like in the diagram.

```
        1    6
   5 | 8    4    9    6

5x1   5
      3    4
     -
5x6   3    0
           4    9
```

Step 3 – How many 5's go into 34? It is 6 whole which is placed at the top of 4.

Step 4 – The answer 6 is then multiplied by the divisor 5 (5 x 6). The answer 30 is then placed under the 34. Then underline it like in the diagram. Now subtract 30 from 34.

The answer is 4 which is directly placed under the 0 (below the line like in the diagram) Now bring down the next number (9) to make the number 49 as shown in the diagram.

```
        1    6    9
   5 | 8    4    9    6

5x1   5
      3    4
     -
5x6   3    0
           4    9
          -
5x9        4    5
                4
```

Step 5 – How many 5's go into 49? It is 9 whole which is placed at the top of 9.

Step 6 – The answer 9 is then multiplied by the divisor 5 (5 x 9). The answer 45 is then placed under the 49. Then underline it like in the diagram. Now subtract 45 from 49.

The answer is 4 which is directly placed under the 5 (below the underline like in the diagram) Now bring down the next number (6) to make the number 46 as shown in the diagram.

```
        1    6    9    9   r 1
   5 | 8    4    9    6

5x5   5
     -
      3    4
5x6   3    0
           4    9
          -
5x9        4    5
                4    6
               -
5x9             4    5
                     1
```

Step 7 – How many 5's go into 46? It is 9 whole which is placed at the top of 6.

Step 8 – The answer 9 is then multiplied by the divisor 5 (5 x 9). The answer 45 is then placed under the 46 then underline it like in the diagram. Now subtract 45 from 46.

The answer is 1 which is directly placed under the 5 (below the underline like in the diagram).

The remainder is 1 which is placed beside the answer.

Long Division - Without remainders

Calculate 650 ÷ 26

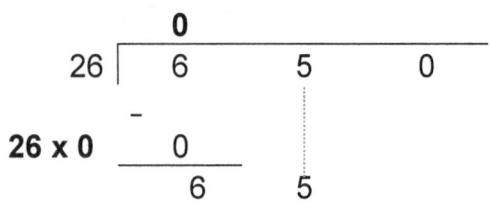

```
          0
   26 | 6   5   0
        -
 26 x 0   0
        _____
          6   5
```

Step 1 – How many 26's go into 6? It is 0. This is placed on top of the 6.

Step 2 – The 0 is multiplied by 26 (divisor) and the answer 0 is placed under the 6 this time and underlined.

Step 3 – Now subtract 0 from 6. The answer is placed under the underline.

Step 4 - Bring down the next digit which is 5 to make the number 65 (as shown in the diagram).

```
          0   2
   26 | 6   5   0
        -
 26 x 0   0
        _____
          6   5
        -
 26 x 2   5   2
        _____
          1   3   0
```

Step 5 – Divide the number 65 by 26 = 2. This is placed on top of 5. Ignore any remainders at this point.

Step 6 – The answer 2 is multiplied by 26 (divisor). The answer is 52 which is placed under 65 and underlined. Step 7 – now subtract 52 from 65. The answer is 13.

Step 8 - Bring down the next digit which is 0 to make the number 130 (as shown in the diagram).

```
          0   2   5
   26 | 6   5   0
        -
 26 x 0   0
        _____
          6   5
        -
 26 x 2   5   2
        _____
          1   3   0
        -
 26 x 5   1   3   0
        _____
                  0
```

Step 9 – Divide the number 130 by 26 = 5. This is placed on top of 5. Ignore any remainders at this point.

Step 10 – The answer 5 is multiplied by 26 (divisor). The answer is 130 which is placed under 130 and underlined.

Step 11 – now subtract 130 from 130. The answer is 0.

There are no remainders - 650 ÷ 26 = 25

Long Division - With remainders

Share 16 into 3965

```
            0
    16 | 3   9   6   5

16 x 0    0
          3   9
```

Step 1 – How many 6's go into 3? It is 0. This is placed on top of the 3.

Step 2 – The 0 is multiplied by 16 (divisor) and the answer 0 is placed under the 6 this time and underlined.
Step 3 – Now subtract 0 from 3. The answer is placed under the underline.

Step 4 - Bring down the next digit which is 9 to make the number 39 (as shown in the diagram).

```
            0   2
    16 | 3   9   6   5

16 x 0    0
          3   9
16 x 2    3   2
              7   6
```

Step 5 – Divide the number 39 by 16 = 2. This is placed on top of 9. Ignore any remainders at this point.
Step 6 – The answer 2 is multiplied by 16 (divisor) the answer is 32 which is placed under 39 and underlined.

Step 7 – Now subtract 32 from 39. The answer is 7.

Step 8 - Bring down the next digit which is 6 to make the number 76 (as shown in the diagram).

```
            0   2   4
    16 | 3   9   6   5

16 x 0    0
          3   9
16 x 2    3   2
              7   6
16 x 4        6   4
              1   2   5
```

Step 9 – Divide the number 76 by 16 = 4. This is placed on top of 6. Ignore any remainders at this point.
Step 10 – The answer 4 is multiplied by 16 (divisor). The answer is 64 which is placed under 76 and underlined.

Step 11 – Now subtract 64 from 76. The answer is 12.

Step 12 - Bring down the next digit which is 5 to make the number 125 (as shown in in the diagram).

```
          0   2   4   7 R 13
    16 | 3   9   6   5

16 x 0    0
          3   9
16 x 2    3   2
              7   6
16 x 4        6   4
              1   2   5
16 x 7        1   1   2
                  1   3
```

Step 13 – Divide the number 125 by 16 = 7. This is placed on top of 5. Ignore any remainders at this point.

Step 14 – The answer 7 is multiplied by 16 (divisor). The answer is 112 which is placed under 125 and underlined.
Step 15 – Now subtract 112 from 125. The answer is **13**. The remainder is 13

$3965 \div 16 = 247 \; r^{13}$

Converting Remainders into Fractions or Decimals

Divide 15 by 4

$$\begin{array}{r} 0 \\ 4\overline{\smash{)}\,15} \end{array}$$

4 x 0 $\underline{0}$

1

Step 1 – How many 4's go into 1? It is 0, which is placed at the top of 1, at this point any remainders are ignored.

Step 2 – The 0 is then multiplied by the 4 (divisor). The answer 0 is then placed under the 1 then underline it like in the diagram.

Now subtract 0 from 1. The answer is 1.

Now bring down the next number (5) to make the number 15 as shown in the diagram.

$$\begin{array}{r} 03 \quad R^3 \\ 4\overline{\smash{)}\,15} \end{array}$$

4 x 0 $\underline{0}$

1   5

4 x3 $\underline{12}$

3

Step 3 – How many 4's go into 15? It is 3 whole which is placed at the top of 5.

Step 4 – The answer 3 is then multiplied by the divisor 4. The answer 12 is then placed under the 15, then underline it as shown the diagram. Now subtract 12 from 15.

The remainder is 3 (see below on how to convert remainders into fractions or decimals).

Converting the remainder into a fraction. So:

$$15 \div 4 = 3 \, \frac{\text{remainder } 3}{\text{divisor } 4}$$

$$3\tfrac{3}{4}$$

Converting into a Decimal

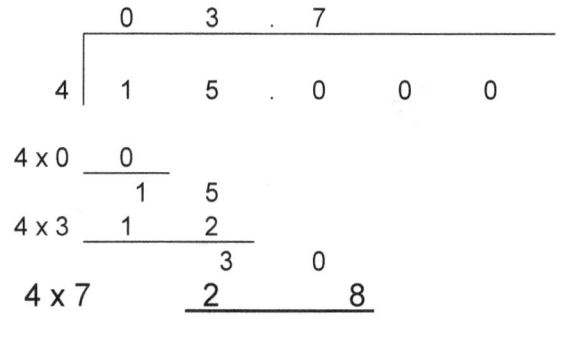

```
        0   3 . 7
   ┌─────────────────────────────
  4│ 1   5 . 0   0   0
4 x 0   0
        1   5
4 x 3   1   2
            3   0
  4 x 7     2       8
```

```
        0   3 . 7   5
   ┌─────────────────────
  4│ 1   5 . 0   0
4 x 0   0
        1   5
4 x3    1   2
            3   0
4 x7        2   8
                2   0
```

```
4 x 5           2   0
                    0
```

Converting Remainders to Decimals.

Step 1- Add in a decimal point as shown in the diagram, then add in some zeros.

Step 2 – Bring down the 0 to make the number 30. Step 3 –

Divide 30 by 4 = 7.

Ignore remainders at this point.

Place the 7 at the top of the 0 like in the diagram. Step 4 –

Multiply 7 by 4 = 28, place this under the 30.

Step 5 – Bring the 0 down to make the number 20.

Step 6 – Divide 20 by 4 = 5 place this beside the 7

Step 7 – Final step: multiply 4 by 5 = 20. Place this under the 20 then subtract = 0. This is where you stop.

So: $15 \div 4 = 3.75$

Exercise D8 - Find the quotient (Without remainders)

$799\overline{\smash{)}8789}$ $319\overline{\smash{)}6380}$ $194\overline{\smash{)}3880}$ $495\overline{\smash{)}8415}$ $595\overline{\smash{)}1190}$

$63\overline{\smash{)}567}$ $18\overline{\smash{)}108}$ $22\overline{\smash{)}352}$ $11\overline{\smash{)}396}$ $11\overline{\smash{)}385}$

$52\overline{\smash{)}416}$ $357\overline{\smash{)}7497}$ $254\overline{\smash{)}6350}$ $37\overline{\smash{)}481}$ $326\overline{\smash{)}1956}$

$401\overline{\smash{)}1604}$ $51\overline{\smash{)}714}$ $72\overline{\smash{)}360}$ $94\overline{\smash{)}470}$ $46\overline{\smash{)}322}$

$32\overline{\smash{)}800}$ $23\overline{\smash{)}621}$ $24\overline{\smash{)}480}$ $84\overline{\smash{)}504}$ $59\overline{\smash{)}354}$

Working out box

Exercise D9 - Find the quotient (Without remainders)

$942\overline{)942}$ $491\overline{)982}$ $585\overline{)5850}$ $301\overline{)8428}$ $421\overline{)5473}$

$666\overline{)9324}$ $732\overline{)3660}$ $544\overline{)7072}$ $390\overline{)5070}$ $759\overline{)6831}$

$153\overline{)5049}$ $437\overline{)8303}$ $552\overline{)8280}$ $821\overline{)821}$ $302\overline{)6644}$

$369\overline{)1476}$ $305\overline{)9760}$ $950\overline{)7600}$ $524\overline{)7860}$ $473\overline{)7568}$

Working out box

Exercise D10 - Find the quotient (With remainders)

$87 \overline{)\ 156}$ $74 \overline{)\ 851}$ $28 \overline{)\ 825}$ $31 \overline{)\ 798}$ $90 \overline{)\ 570}$

$20 \overline{)\ 336}$ $89 \overline{)\ 103}$ $37 \overline{)\ 378}$ $94 \overline{)\ 190}$ $91 \overline{)\ 463}$

$77 \overline{)\ 325}$ $39 \overline{)\ 963}$ $26 \overline{)\ 451}$ $55 \overline{)\ 905}$ $20 \overline{)\ 658}$

$12 \overline{)\ 860}$ $12 \overline{)\ 507}$ $42 \overline{)\ 296}$ $17 \overline{)\ 691}$ $57 \overline{)\ 329}$

Working out box

Exercise D11 - Find the quotient (Without remainders)

$697)\overline{2091}$ $240)\overline{9600}$ $142)\overline{3692}$ $677)\overline{6093}$ $102)\overline{9486}$

$21)\overline{8631}$ $272)\overline{3808}$ $329)\overline{5922}$ $464)\overline{2784}$ $435)\overline{8700}$

$542)\overline{7046}$ $239)\overline{5497}$ $434)\overline{2604}$ $35)\overline{6160}$ $960)\overline{9600}$

$72)\overline{3960}$ $877)\overline{2631}$ $978)\overline{4890}$ $482)\overline{9158}$ $377)\overline{4524}$

Working out box

Exercise D12 - Find the quotient (With remainders)

$69\overline{)416}$ $54\overline{)708}$ $89\overline{)975}$ $26\overline{)527}$ $32\overline{)218}$

$27\overline{)672}$ $98\overline{)403}$ $35\overline{)170}$ $82\overline{)860}$ $55\overline{)250}$

$28\overline{)652}$ $18\overline{)958}$ $56\overline{)242}$ $86\overline{)738}$ $35\overline{)861}$

$42\overline{)156}$ $79\overline{)406}$ $47\overline{)277}$ $11\overline{)178}$

Working out box

Exercise D13 - Find the quotient (With remainders)

$62{\overline{\smash{)}\,686}}$ $31{\overline{\smash{)}\,477}}$ $95{\overline{\smash{)}\,363}}$ $34{\overline{\smash{)}\,726}}$ $65{\overline{\smash{)}\,824}}$

$72{\overline{\smash{)}\,295}}$ $45{\overline{\smash{)}\,550}}$ $86{\overline{\smash{)}\,458}}$ $73{\overline{\smash{)}\,325}}$ $58{\overline{\smash{)}\,765}}$

$20{\overline{\smash{)}\,559}}$ $35{\overline{\smash{)}\,337}}$ $61{\overline{\smash{)}\,888}}$ $72{\overline{\smash{)}\,451}}$ $97{\overline{\smash{)}\,912}}$

$57{\overline{\smash{)}\,486}}$ $37{\overline{\smash{)}\,363}}$ $59{\overline{\smash{)}\,576}}$ $57{\overline{\smash{)}\,852}}$ $71{\overline{\smash{)}\,992}}$

Working out box

Exercise P1

a) seven thousand two hundred eighty-four

b) two hundred eighty-four

c) seven thousand five hundred thirty-five

d) one thousand four hundred eighty-seven

e) six hundred forty-two

f) twenty-three

g) seven hundred fifteen

h) seventy-eight

i) four thousand eight hundred eighty-eight

j) forty-three

Exercise P2

a) 6213
b) 8704
c) 4742
d) 5917
e) 803
f) 9023
g) 9274
h) 2329
i) 8326
j) 8988

Exercise P3

a) 9395
b) 1010
c) 966
d) 6144
e) 3666
f) 6794
g) 6620
h) 7838
i) 7576
j) 9250

Exercise P4

a) 8 tens
b) 5 tens
c) 5 tens
d) 7 units
e) 4 tens
f) 9 tens
g) 8 hundreds
h) 0 units
i) 4 hunreds
j) 2 tens
k) 6 thousands
l) 0 hundreds

Exercise P5

a) 0 tens
b) 6 units
c) 2 tens
d) 1 unit
e) 9 tens
f) 0 tens
g) 6 tenths
h) 2 tens
i) 8 tenths
j) 3 units
k) 7 tenths
l) 7 units

Exercise P 6

a) 5300   b) 6770   c) 10000   d) 6000   e) 4800   f) 5100   g) 2610   h) 8700   i) 6300   j) 6700

k) 17300   l) 67210

Exercise P7

a)	b)	c)	d)	e)	f)	g)	h)	i)	j)
673	749	634	984	(254)	683	(303)	729	(510)	290
894	473	692	(280)	767	556	526	479	888	526
318	(312)	(575)	803	451	514	663	(116)	774	(261)
(258)	338	597	461	770	(246)	582	545	651	336

Exercise P8

a)	b)	c)	d)	e)	f)	g)	h)	i)	j)
277	182	308	305	233	157	183	341	500	249
513	529	369	370	273	454	228	428	573	453
770	682	665	854	284	507	426	658	587	565
771	963	910	972	655	831	890	661	625	838

Exercise P9

a)	b)	c)	d)	e)	f)	g)	h)	i)	j)
9.07	3.18	39.9	6.06	41.0	3.22	59.8	4.49	5.70	7.28
45.4	49.2	654	9.35	355	3.34	71.2	42.5	31.3	8.38
57.9	488	791	103	422	351	378	117	59.2	241
134	954	869	747	769	807	445	151	64.7	334

Exercise P10

a) 3907 > 3   b) 307 < 6065   c) 8 > 4   d) 3 < 4198   e) 77 > 2

f) 13 < 481   g) 3 < 73   h) 5814 > 4   i) 488 > 4   j) 15 < 27

Exercise A 1

a.	110 + 155 =	D		J = 280
b.	61 + 14 =	A		B = 272
c.	103 + 21 =	G		I = 169
d.	157 + 53 =	F		C = 252
e.	176 + 104 =	J		E = 311
f.	199 + 112 =	E		D = 265
g.	98 + 71 =	I		A = 75
h.	18 + 93 =	H		H = 111
i.	70 + 182 =	C		G = 124
j.	69 + 203 =	B		F = 210

Exercise A 2

$4 + \underline{15} = 19$ $10 + \underline{6} = 16$ $6 + \underline{4} = 10$ $11 + \underline{7} = 18$

$12 + \underline{0} = 12$ $12 + \underline{5} = 17$ $3 + \underline{1} = 4$ $7 + \underline{3} = 10$

$2 + \underline{0} = 2$ $3 + \underline{4} = 7$ $15 + \underline{3} = 18$ $5 + \underline{4} = 9$

    $6 + \underline{6} = 12$ $2 + \underline{8} = 10$ $5 + \underline{10} = 15$

$10 + \underline{7} = 17$ $16 + \underline{2} = 18$ $3 + \underline{2} = 5$

$3 + \underline{3} = 6$ $2 + \underline{1} = 3$

Exercise A3

60	72	36	17	13	58	84	43	82
+ 62	+ 74	+ 38	+ 19	+ 15	+ 60	+ 86	+ 45	+ 84
122	146	74	36	28	118	170	88	166

59	52	68	64	89	91	35	80	6
+ 61	+ 54	+ 70	+ 66	+ 91	+ 93	+ 37	+ 82	+ 8
120	106	138	130	180	184	72	162	14

45	10	55	21	87	63	11	27	30
+ 47	+ 12	+ 57	+ 23	+ 89	+ 65	+ 13	+ 29	+ 32
92	22	112	44	176	128	24	56	62

Exercise A4

```
    61       50       26       92       42       53       98       95       33       26
    34       64       55       38       85       98       92       15       27       90
  + 40     + 14     + 11     + 18     + 30     + 41     + 64     + 24     + 15     + 76
  -----    -----    -----    -----    -----    -----    -----    -----    -----    -----
   135      128       92      148      157      192      254      134       75      192
```

Exercise A5

```
  503       503      143      430      956      431      186      373
+  51     +  26    +  46    +  51    +  30    +  30    +  12    +  16
-----     -----    -----    -----    -----    -----    -----    -----
  554       529      189      481      986      461      198      389

  333       783      389      772      223      440      981      124
+  52     +  11    +  10    +  16    +  56    +  26    +  12    +  60
-----     -----    -----    -----    -----    -----    -----    -----
  385       794      399      788      279      466      993      184

  150       433      421      352      875      401      363      108
+  33     +  13    +  36    +  32    +  10    +  36    +  26    +  30
-----     -----    -----    -----    -----    -----    -----    -----
  183       446      457      384      885      437      389      138

  323       803      189      802      400      246
+  56     +  96    +  10    +  96    +  96    +  30
-----     -----    -----    -----    -----    -----
  379       899      199      898      496      276
```

Exercise A6

```
  236      356      965      360      230      538      837      785       86       97
   98       62      231      562      398      389      939       37      962      662
+  56     + 70    +  51    + 13    + 61    + 32    + 83    + 816   + 721    + 972
-----     -----    -----    -----    -----    -----    -----    -----    -----    -----
  390      488     1247      935      689      959     1859     1638     1769     1731
```

Exercise A7

```
  550      133      624      494      791      174      685      872      704      428
+ 332    + 643    + 244    + 402    + 200    + 524    + 201    + 125    + 153    + 420
-----    -----    -----    -----    -----    -----    -----    -----    -----    -----
  882      776      868      896      991      698      886      997      857      848

  240      453      479      805      206      208      484      554      769      314
+ 643    + 223    + 510    + 111    + 363    + 611    + 201    + 145    + 220    + 231
-----    -----    -----    -----    -----    -----    -----    -----    -----    -----
  883      676      989      916      569      819      685      699      989      545

  846      652      560      643      306      206      559      435      113      260
+ 121    + 224    + 118    + 114    + 542    + 172    + 310    + 312    + 441    + 417
-----    -----    -----    -----    -----    -----    -----    -----    -----    -----
  967      876      678      757      848      378      869      747      554      677
```

Exercise A8

272	703	176	984	856	121	751	757	186	904
321	972	930	878	406	302	164	868	625	139
+ 280	+ 682	+ 439	+ 497	+ 715	+ 912	+ 386	+ 519	+ 318	+ 807
873	2357	1545	2359	1977	1335	1301	2144	1129	1850

Exercise A9

5450	9313	4285	8801	9803	4011	2260	9107
+ 217	+ 255	+ 601	+ 130	+ 161	+ 853	+ 228	+ 280
5667	9568	4886	8931	9964	4864	2488	9387

2296	8715	2390	2870	8441	9469	6632	6674
+ 301	+ 100	+ 402	+ 114	+ 233	+ 320	+ 267	+ 215
2597	8815	2792	2984	8674	9789	6899	6889

7122	5276	5167	7432	4004	5897	3119	1340
+ 322	+ 611	+ 712	+ 300	+ 500	+ 101	+ 710	+ 153
7444	5887	5879	7732	4504	5998	3829	1493

8593	3510	4246	6804	8105	8679
+ 303	+ 143	+ 640	+ 143	+ 542	+ 210
8896	3653	4886	6947	8647	8889

Exercise A10

		986			3290		1222
4697	1802	8038	3566	5440	5966	351	955
1351	4865	1306	6125	181	1029	1437	7684
+ 4806	+ 9290	+ 5937	+ 7502	+ 9035	+ 6638	+ 9313	+ 7931
10854	15957	16267	17193	14656	16923	11101	17792

	6250	9970	3169	2853		7087	1911
8,002	3498	1131	2336	225	8765	2154	2352
252	6239	9166	7970	3406	9577	2639	3368
+ 7262	+ 7468	+ 770	+ 2111	+ 6347	+ 7495	+ 2657	+ 2031
15516	23455	21037	15586	12831	25837	14537	9662

3692		9680	
956	2978	177	5462
7174	7895	9879	9043
+ 1468	+ 4037	+ 143	+ 9862
13290	14910	19879	24367

83

Exercise A11

214 + 3115 = 3329	2724 + 3103 = 5827	6307 + 2451 = 8758	137 + 1031 = 1168	8814 + 1061 = 9875	6040 + 2417 = 8457	6571 + 2010 = 8581	8210 + 1225 = 9435
120 + 1322 = 1442	3540 + 1014 = 4554	1734 + 5115 = 6849	3950 + 1003 = 4953	5302 + 176 = 5478	2504 + 3051 = 5555	7604 + 1024 = 8628	563 + 7213 = 7776
2331 + 1345 = 3676	6565 + 1112 = 7677	5854 + 1043 = 6897	7641 + 158 = 7799	5542 + 2026 = 7568	5853 + 103 = 5956	1165 + 6,223 = 7388	2580 + 4206 = 6786
4819 + 3010 = 7829	4360 + 4602 = 8962	6528 + 310 = 6838	2784 + 5002 = 7786	7235 + 1632 = 8867	3201 + 4067 = 7268		

Exercise A12

	9427			5388		6085
5172	7499	6584	3290	7704	4250	5459
2442	9979	2953	1914	4295	5208	5869
8674	6055	3937	2567	7614	9736	6900
+ 7320	+ 3671	+ 6916	+ 8242	+ 1619	+ 3088	+ 6060
23608	36631	20390	16013	26620	22282	30373
6844			1879		5085	
1878	4474	7335	8678	6391	8481	7580
1143	8079	2888	9796	8670	4652	7437
2210	8518	9680	4631	6379	8189	2514
+ 9907	+ 7725	+ 6267	+ 7871	+ 8095	+ 4565	+ 5620
21982	28796	26170	32855	29535	30972	23151
1285	6872		1935	2553	8745	
6767	2179	6866	7979	7583	6788	
1623	2004	9975	4428	9475	2658	
6105	8089	6956	2157	8226	7555	
+ 3158	+ 1262	+ 5824	+ 3288	+ 9677	+ 2817	
18938	20406	29621	19787	37514	28563	

Exercise A13

82

118

78

39

23

47

215

915

83

483

Exercise A14

10	0	14
12	8	4
2	16	6

Magic Number: 24

10	5	6
3	7	11
8	9	4

Magic Number: 21

12	10	20
22	14	6
8	18	16

Magic Number: 42

Exercise A15

						31
8	5	7	3	7	2	32
3	6	1	6	3	7	26
4	7	7	1	9	3	31
6	5	9	8	8	4	40
3	9	2	3	1	4	22
7	2	2	7	6	1	25
31	34	28	28	34	21	31

Exercise A16 - Solve.

7	+	10	+	10	=	27
+		+		+		+
1	+	3	+	2	=	6
+		+		+		+
9	+	6	+	5	=	20
=		=		=		=
17	+	19	+	17	=	53

1	+	9	+	3	=	13
+		+		+		+
4	+	8	+	4	=	16
+		+		+		+
8	+	1	+	4	=	13
=		=		=		=
13	+	18	+	11	=	42

Exercise A17

Solve.

| | | | | | | | 39 |

							34
8	10	2	9	4	6	6	45
7	6	7	7	9	1	10	47
6	8	7	6	6	9	1	43
3	2	4	3	3	2	7	24
8	8	9	2	3	2	8	40
9	5	4	4	8	3	9	42
4	2	2	6	10	7	5	36
45	41	35	37	43	30	46	35

2	4	10	10	10	2	4	3	45
4	1	5	1	2	1	5	4	23
1	6	5	7	3	8	7	9	46
1	2	1	6	2	7	6	1	26
5	6	4	9	6	2	2	7	41
4	7	1	6	3	3	7	2	33
9	7	5	7	4	5	3	9	49
4	3	3	8	8	3	5	2	36
30	36	34	54	38	31	39	37	28

| | | | | | | | 24 |

| | | | | | | | 60 |

						24
1	2	3	8	9	10	2	35
1	5	6	2	6	8	7	35
8	6	4	4	2	3	1	28
9	5	5	6	8	5	1	39
7	1	3	3	10	9	7	40
7	1	5	1	6	9	1	30
2	8	3	1	8	5	4	31
35	28	29	25	49	49	23	39

1	4	8	5	5	6	10	1	4	7	51
4	8	5	9	6	3	3	4	3	3	48
2	2	2	9	4	8	4	10	4	9	54
7	2	2	6	6	5	8	1	4	7	48
2	1	5	2	5	2	6	6	3	9	41
8	4	3	2	8	2	1	6	1	8	43
10	9	6	4	1	6	9	2	6	6	59
3	6	6	6	1	1	1	2	6	2	34
4	8	5	7	8	2	7	5	6	2	54
4	2	9	4	5	2	2	3	6	10	47
45	46	51	54	49	37	51	40	43	63	51

Exercise A20

8.4	5.2	0.30	7.7	0.16	0.16	1.4	7.4	0.96	8.3
+ 3.1	+ 1.2	+ 0.54	+ 4.1	+ 0.26	+ 0.27	+ 1.0	+ 5.4	+ 0.25	+ 4.0
11.5	6.4	0.84	11.8	0.42	0.43	2.4	12.8	1.21	12.3

Exersise A21

0.079	65	0.029	34	36	0.088	0.014
+ 4.955	+ 2,785	+ 1.504	+ 658	+ 3,005	+ 1.969	+ 4.993
5.034	2,850	1.533	692	3,041	2.057	5.007

13	0.026	76
+ 2955	+ 4.572	+ 3041
2968	4.598	3117

Exercise A22

1.7	96	0.043	0.051	0.085	7.5
+ 9.5	+ 3,208	+ 0.960	+ 1.763	+ 3.290	+ 101.2
11.2	3,304	1.003	1.814	3.375	108.7

4.2	0.52	3.4
+ 121.6	+ 49.42	+ 193.4
125.8	49.94	196.8

Exercise S1

20	40	14	44	50	18	17	15	12
- 11	- 30	- 7	- 35	- 41	- 9	- 9	- 7	- 4
9	10	7	9	9	9	8	8	8

21	21	15	24	19	14	22	12	12
- 11	- 12	- 9	- 14	- 9	- 6	- 12	- 8	- 7
10	9	6	10	10	8	10	4	5

14	22
- 9	- 13
5	9

Exercise S2

57	45	25	76	88	44	22	82	28	28
- 45	- 15	- 17	- 38	- 73	- 33	- 15	- 63	- 14	- 17
12	30	8	38	15	11	7	19	14	11

41	68	67	27	41	79	75	58	44	80
- 11	- 48	- 50	- 11	- 29	- 65	- 32	- 44	- 31	- 65
30	20	17	16	12	14	43	14	13	15

51	27	73	61	82	84	80	39	54	47
- 31	- 18	- 65	- 54	- 74	- 14	- 40	- 25	- 23	- 34
20	9	8	7	8	70	40	14	31	13

Exercise S3

46	207	97	294	720	610	94	294	846
- 36	- 103	- 87	- 106	- 615	- 515	- 84	- 146	- 212
10	104	10	188	105	95	10	148	634

611	210	953	978	200	45	717	922	727
- 110	- 100	- 256	- 805	- 109	- 35	- 400	- 127	- 427
501	110	697	173	91	10	317	795	300

Exercise S4

361	159	547	461	184	552	295	871	709	559
+ -349	+ -69	+ -433	+ -356	+ -87	+ -494	+ -237	+ -840	+ -671	+ -443
12	90	114	105	97	58	58	31	38	116

603	594	520	211	238	804	184	466	344	259
+ -525	+ -572	+ -408	+ -165	+ -155	+ -747	+ -84	+ -439	+ -230	+ -161
78	22	112	46	83	57	100	27	114	98

Exercise S5

a.	155 - 131 =	J	A = 18
b.	136 - 25 =	E	G = 130
c.	84 - 46 =	H	I = 0
d.	27 - 21 =	B	C = 66
e.	127 - 78 =	D	B = 6
f.	186 - 56 =	G	E = 111
g.	153 - 87 =	C	H = 38
h.	183 - 13 =	F	D = 49
i.	38 - 20 =	A	J = 24
j.	87 - 87 =	I	F = 170

Exercise S6

1108 - 193 915	2804 - 103 2701	5504 - 511 4993	6803 - 632 6171	8209 - 147 8062	3401 - 251 3150	9505 - 400 9105
7706 - 441 7265	2705 - 193 2512	4907 - 582 4325	3302 - 341 2961	9101 - 161 8940	3402 - 252 3150	303 - 123 180
5508 - 175 5333	1401 - 490 911	4104 - 134 3970	1800 - 410 1390	3901 - 591 3310	5004 - 221 4783	

Exercise S7

34000 - 926 33074	56000 - 159 55841	63000 - 340 62660	5000 - 464 4536	40000 - 280 39720	27000 - 210 26790
39000 - 433 38567	2000 - 242 1758	17000 - 876 16124	8000 - 801 7199	17000 - 142 16858	6000 - 118 5882
8000 - 736 7264	65000 - 530 64470	5000 - 776 4224	48000 - 958 47042	30000 - 230 29770	87000 - 401 86599
3000 - 243 2757	98000 - 488 97512				

Exercise S8

773

711

726

254

25

78

813

64

504

287

Exercise S9

7.4	0.29	0.081	0.080	3.0	0.48	9.7
- 4.3	- 0.18	- 0.018	- 0.017	- 1.2	- 0.45	- 7.9
3.1	0.11	0.063	0.063	1.8	0.03	1.8

0.082	49	0.081	0.037	0.97	0.080	0.056
- 0.078	- 27	- 0.072	- 0.025	- 0.85	- 0.061	- 0.044
0.004	22	0.009	0.012	0.12	0.019	0.012

9.8	0.79	7.1	0.044	29	3.5	0.68
- 4.7	- 0.66	- 5.7	- 0.039	- 14	- 1.3	- 0.63
5.1	0.13	1.4	0.005	15	2.2	0.05

70	9.3	9.6	88	6.3	80	0.046
- 45	- 4.2	- 7.12	- 79	- 3.6	- 45	- 0.011
25	5.1	2.48	9	2.7	35	0.035

0.47	79
- 0.25	- 72
0.22	7

Exercise S10

69 - 17 = 52   69 - 0.013 = 68.987   0.036 - 0.033 = 0.003

8.1 - 0.075 = 8.025   0.092 - 0.034 = 0.058   98 - 4.5 = 93.5

91 - 3.4 = 87.6   0.62 - 0.46 = 0.16   0.037 - 0.025 = 0.012

0.065 - 0.045 = 0.020   0.68 - 0.26 = 0.42   85 - 11 = 74

0.037 - 0.016 = 0.021   73 - 33 = 40   83 - 2.5 = 80.5

0.068 - 0.063 = 0.005   0.24 - 0.013 = 0.227   0.096 - 0.063 = 0.033

64 - 1.8 = 62.2   0.82 - 0.69 = 0.13

Exercise S11

47	-	14	-	19	=	14
-		-		-		-
15	-	1	-	8	=	6
-		-		-		-
20	-	5	-	8	=	7
=		=		=		=
12	-	8	-	3	=	1

46	-	15	-	17	=	14
-		-		-		-
11	-	4	-	6	=	1
-		-		-		-
17	-	8	-	2	=	7
=		=		=		=
18	-	3	-	9	=	6

3	-	2	+	5	=	6
-		+		-		+
2	+	1	-	1	=	2
+		-		+		+
4	-	1	+	4	=	7
=		=		=		=
5	+	2	+	8	=	15

9	-	8	+	3	=	4
-		+		-		+
8	+	6	-	1	=	13
+		-		+		+
9	-	1	+	10	=	18
=		=		=		=
10	+	13	+	12	=	35

The secret trail.

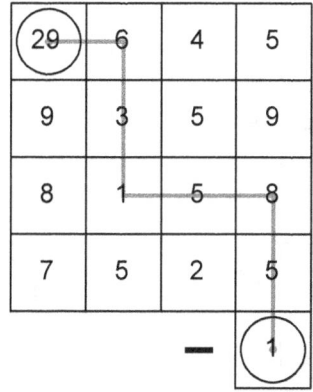

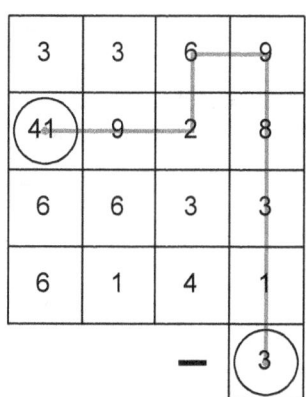

Exercise M 1

8× 7 = 56   10 × 5 = 50   13 × 5 = 65   12 × 5= 60

10 × 7= 70   12 × 3 = 36   10× 9 = 90   14 × 8 = 112

8× 6= 48   11 × 2= 22   8 × 5 = 40   11 × 3= 33

12 × 6 = 72   9 × 7 = 63   15× 4 = 60   14 × 4 = 56

10 × 2 = 20   11 × 8 = 88   3 × 8 = 24   11 × 7 =77

5 × 8 = 40   13 × 6 = 78   8 × 2 = 16   13 × 8 = 104

Exercise M2

x	8	6	11	10	2
4	x	24	44	40	8
15	120	90	165	150	30
3	24	18	33	30	6
5	40	30	55	50	10
10	80	60	110	100	20

x	7	9	15	8	14
9	63	81	135	72	126
12	84	108	180	96	168
10	70	90	150	80	140
6	42	54	90	48	84
8	56	72	120	64	112

x	9	6	8	14	5
2	18	12	16	28	10
11	99	66	88	154	55
12	108	72	96	168	60
8	72	48	64	112	40
13	117	78	104	182	65

32	9	6	12	7	5
6	54	36	72	42	30
9	81	54	108	63	45
4	36	24	48	28	20
12	108	72	144	84	60
3	27	18	36	21	15

Exercise M3

38 × 2 76	66 × 4 264	22 × 8 176	23 × 4 92	69 × 5 345	4 × 4 16	78 × 8 624	14 × 4 56
69 × 3 207	62 × 5 310	87 × 2 174	9 × 5 45	89 × 8 712	89 × 5 445	12 × 7 84	59 × 3 177
16 × 9 144	34 × 9 306	6 × 8 48	88 × 8 704				

Exercise M4

784 × 5 3920	690 × 7 4830	649 × 8 5192	609 × 4 2436	168 × 8 1344	303 × 9 2727	681 × 7 4767
906 × 8 7248	473 × 5 2365	846 × 6 5076	351 × 7 2457	708 × 9 6372	869 × 5 4345	811 × 7 5677
947 × 8 7576	211 × 5 1055	671 × 7 4697	547 × 6 3282	338 × 6 2028	648 × 6 3888	

Exercise M5

3200 × 3 9600	1322 × 2 2644	1010 × 4 4040	2201 × 6 13206	2332 × 3 6996	2212 × 4 8848	1340 × 8 10720
1121 × 4 4484	1100 × 4 4400	2404 × 5 12020	4141 × 7 28987	1120 × 4 4480	1256 × 4 5024	2120 × 4 8480
1123 × 6 6738	3301 × 3 9903	4344 × 8 34752	1313 × 3 3939	1110 × 5 5550	6809 × 9 61281	

Exercise M6

11	35	9	92	72	25	43	31
× 11	× 11	× 40	× 83	× 80	× 89	× 31	× 76
121	385	360	7636	5760	2225	1333	2356

35	49	20	39	47	1	48	16
× 53	× 32	× 30	× 67	× 80	× 10	× 37	× 67
1855	1568	600	2613	3760	10	1776	1072

63	44	10	85
× 73	× 69	× 13	× 60
4599	3036	130	5100

Exercise M7

41	61	97	15	10	43	28	82	65
× 42	× 61	× 96	× 87	× 89	× 50	× 61	× 78	× 24
1722	3721	9312	1305	890	2150	1708	6396	1560

39	64	416	661	843	848	405	166
× 92	× 51	× 85	× 17	× 19	× 19	× 63	× 21
3588	3264	35360	11237	16017	16112	25515	3486

174	143	566
× 69	× 83	× 16
12006	11869	9056

Exercise M8

760	412	348	314	699	226	677	324
× 220	× 249	× 327	× 762	× 416	× 793	× 552	× 28
167200	102588	113796	239268	290784	179218	373704	9072

752	247	3010	3204	2314	2222	2111
× 150	× 432	× 23	× 52	× 82	× 36	× 47
112800	106704	69230	166608	189748	79992	99217

1,010	2225	2001	2112	2012
× 85	× 74	× 94	× 99	× 59
85850	164650	188094	209088	118708

Exercise M9

472	226	326	271	502	520	883	745
× 795	× 724	× 760	× 394	× 404	× 651	× 104	× 300
375240	163624	247760	106774	202808	338520	91832	223500

177	162
× 855	× 687
151335	111294

Exercise M10

28 × 0.007 = 0.196

74 × 0.6 = 44.4

47 × 0.6 = 28.2

28 × 0.5 = 14.0

20 × 0.002 = 0.040

39 × 0.006 = 0.234

35 × 0.4 = 14.0

88 × 0.004 = 0.352

43 × 0.2 = 8.6

52 × 0.06 = 3.12

96 × 0.8 = 76.8

13 × 0.8 = 10.4

46 × 0.2 = 9.2

45 × 0.07 = 3.15

65 × 0.09 = 5.85

20 × 0.001 = 0.020

88 × 0.09 = 7.92

98 × 0.008 = 0.784

72 × 0.001 = 0.072

74 × 0.002 = 0.148

28 × 0.6 = 16.8

88 × 0.003 = 0.264

44 × 0.8 = 35.2

30 × 0.004 = 0.120

54 × 0.02 = 1.08

Exercise M12 -

53 × 40 = 2120

70 × 20 = 1400

93 × 80 = 7440

59 × 20 = 1180

60 × 30 = 1800

66 × 30 = 1980

48 × 40 = 1920

29 × 60 = 1740

61 × 60 = 3660

65 × 40 = 2600

37 × 60 = 2220

14 × 20 = 280

62 × 40 = 2480

35 × 80 = 2800

79 × 10 = 790

71 × 70 = 4970

94 × 40 = 3760

45 × 90 = 4050

32 × 60 = 1920

28 × 50 = 1400

Exercise M11

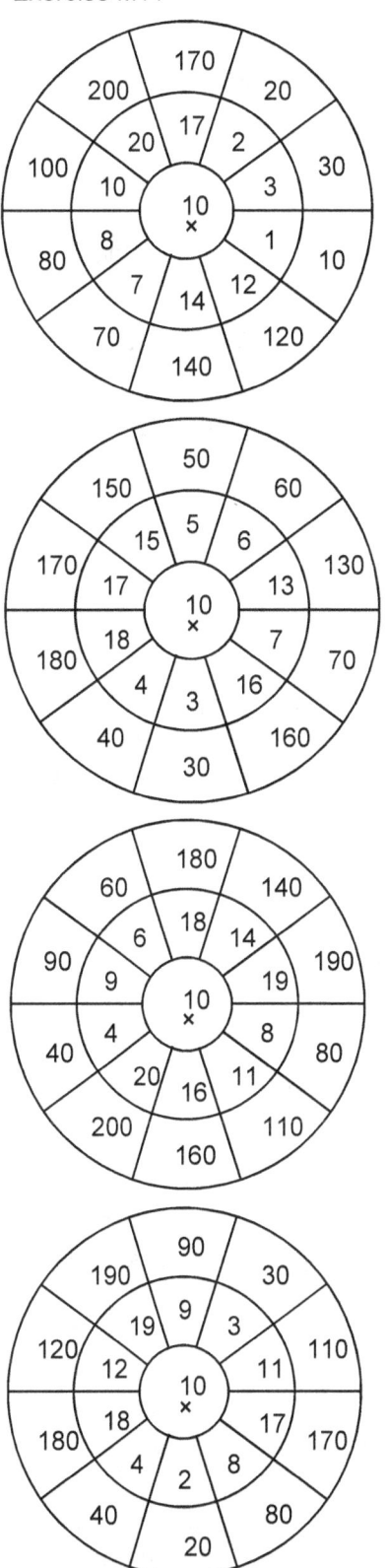

Exercise M13

27 × 300 = 8100   61 × 500 = 30500   90 × 500 = 45000

72 × 300 = 21600   18 × 500 = 9000   78 × 300 = 23400

67 × 200 = 13400   59 × 500 = 29500   93 × 500 = 46500

51 × 900 = 45900   12 × 600 = 7200   80 × 700 = 56000

99 × 200 = 19800   92 × 300 = 27600   88 × 500 = 44000

12 × 300 = 3600   65 × 900 = 58500   47 × 900 = 42300

83 × 100 = 8300   96 × 400 = 38400

Exercise M14

18 × 2,000 = 36000   73 × 2000 = 146000   42 × 8000 = 336000

84 × 8,000 = 672000
                         22 × 6000 = 132000   33 × 4000 = 132000

60 × 2000 = 120000
                         71 × 9000 = 639000   69 × 2000 = 138,000

18 × 3000 = 54000
                         94 × 9000 = 846000   55 × 2000 = 110000

58 × 9000 = 522000
                         20 × 1000 = 20000   29 × 7000 = 203000

93 × 9000 = 837000
                         32 × 9000 = 288000   71 × 3000 = 213000

21 × 4000 = 84000
                         46 × 6000 = 276000

Exercise D1

333 ÷ 10 = 33.3   175 ÷ 10 = 17.5   529 ÷ 10 = 52.9   676 ÷ 10 = 67.6

807 ÷ 10 = 80.7   202 ÷ 10 = 20.2   892 ÷ 10 = 89.2   297 ÷ 10 = 29.7

499 ÷ 10 = 49.9   205 ÷ 10 = 20.5

Exercise D2

808 ÷ 100 = 8.1   470 ÷ 100 = 4.7   720 ÷ 100 = 7.2   176 ÷ 100 = 1.8

953 ÷ 100 = 9.5   770 ÷ 100 = 7.7   463 ÷ 100 = 4.6   652 ÷ 100 = 6.5

441 ÷ 100 = 4.4   622 ÷ 100 = 6.2

Exercise D4

927 ÷ 1000 = 0.9   727 ÷ 1000 = 0.7   732 ÷ 1000 = 0.7

101 ÷ 1000 = 0.1   206 ÷ 1000 = 0.2   496 ÷ 1000 = 0.5

960 ÷ 1000 = 1.0   554 ÷ 1000 = 0.6   513 ÷ 1000 = 0.5

259 ÷ 1000 = 0.3

Exercise D4

28 ÷ 7 = 4   20 ÷ 5 = 4   12 ÷ 6 = 2   48 ÷ 6 = 8   72 ÷ 8 = 9

56 ÷ 8 = 7   12 ÷ 3 = 4   24 ÷ 3 = 8   40 ÷ 8 = 5   25 ÷ 5 = 5

18 ÷ 6 = 3   6 ÷ 2 = 3   49 ÷ 7 = 7   9 ÷ 3 = 3   24 ÷ 4 = 6

15 ÷ 5 = 3   35 ÷ 5 = 7   64 ÷ 8 = 8   36 ÷ 6 = 6   56 ÷ 7 = 8

9 ÷ 9 = 1   16 ÷ 8 = 2   24 ÷ 8 = 3   21 ÷ 3 = 7   45 ÷ 9 = 5

30 ÷ 5 = 6   4 ÷ 4 = 1   6 ÷ 6 = 1   42 ÷ 6 = 7   14 ÷ 7 = 2

63 ÷ 7 = 9   30 ÷ 6 = 5   32 ÷ 8 = 4   7 ÷ 1 = 7   12 ÷ 4 = 3

42 ÷ 7 = 6   8 ÷ 4 = 2   16 ÷ 2 = 8   15 ÷ 3 = 5   18 ÷ 2 = 9

Exercise D5

÷	61	79	31	22	86
3	20 r1	26 r1	10 r1	7 r1	28 r2
7	8 r5	11 r2	4 r3	3 r1	12 r2
5	12 r1	15 r4	6 r1	4 r2	17 r1
4	15 r1	19 r3	7 r3	5 r2	21 r2
9	6 r7	8 r7	3 r4	2 r4	9 r5

÷	84	98	80	45	43
6	14	16 r2	13 r2	7 r3	7 r1
2	42	49	40	22 r1	21 r1
1	84	98	80	45	43
4	21	24 r2	20	11 r1	10 r3
7	12	14	11 r3	6 r3	6 r1

Exercise D6

$$5\overline{)90} = 18 \qquad 3\overline{)78} = 26 \qquad 7\overline{)280} = 40 \qquad 8\overline{)72} = 9 \qquad 5\overline{)760} = 152 \qquad 5\overline{)460} = 92 \qquad 7\overline{)938} = 134$$

$$8\overline{)208} = 26 \qquad 9\overline{)513} = 57 \qquad 7\overline{)154} = 22 \qquad 2\overline{)532} = 266 \qquad 9\overline{)18} = 2 \qquad 3\overline{)33} = 11 \qquad 1\overline{)76} = 76$$

$$2\overline{)282} = 141 \qquad 5\overline{)30} = 6 \qquad 6\overline{)84} = 14 \qquad 6\overline{)78} = 13 \qquad 2\overline{)70} = 35 \qquad 5\overline{)980} = 196 \qquad 4\overline{)72} = 18$$

$$8\overline{)64} = 8 \qquad 5\overline{)50} = 10 \qquad 4\overline{)44} = 11 \qquad 6\overline{)966} = 161$$

Exercise D7

$$7\overline{)196} = 28 \qquad 7\overline{)266} = 38 \qquad 8\overline{)200} = 25 \qquad 3\overline{)183} = 61 \qquad 6\overline{)282} = 47 \qquad 5\overline{)235} = 47 \qquad 6\overline{)384} = 64 \qquad 2\overline{)58} = 29$$

$$4\overline{)176} = 44 \qquad 8\overline{)440} = 55 \qquad 3\overline{)273} = 91 \qquad 5\overline{)190} = 38 \qquad 7\overline{)616} = 88 \qquad 7\overline{)238} = 34 \qquad 6\overline{)96} = 16 \qquad 7\overline{)350} = 50$$

$$5\overline{)255} = 51 \qquad 2\overline{)118} = 59 \qquad 7\overline{)133} = 19 \qquad 1\overline{)54} = 54 \qquad 7\overline{)217} = 31 \qquad 1\overline{)87} = 87 \qquad 8\overline{)728} = 91 \qquad 8\overline{)288} = 36$$

$$5\overline{)165} = 33$$

Exercise D8

$$\frac{11}{799 \overline{)8,789}}$$ $$\frac{20}{319 \overline{)6,380}}$$ $$\frac{20}{194 \overline{)3,880}}$$ $$\frac{17}{495 \overline{)8,415}}$$ $$\frac{2}{595 \overline{)1,190}}$$

$$\frac{9}{63 \overline{)567}}$$ $$\frac{6}{18 \overline{)108}}$$ $$\frac{16}{22 \overline{)352}}$$ $$\frac{35}{11 \overline{)385}}$$ $$\frac{36}{11 \overline{)396}}$$

$$\frac{8}{52 \overline{)416}}$$ $$\frac{21}{357 \overline{)7,497}}$$ $$\frac{25}{254 \overline{)6,350}}$$ $$\frac{13}{37 \overline{)481}}$$ $$\frac{6}{326 \overline{)1,956}}$$

$$\frac{4}{401 \overline{)1,604}}$$ $$\frac{14}{51 \overline{)714}}$$ $$\frac{5}{72 \overline{)360}}$$ $$\frac{5}{94 \overline{)470}}$$ $$\frac{7}{46 \overline{)322}}$$

$$\frac{25}{32 \overline{)800}}$$ $$\frac{27}{23 \overline{)621}}$$ $$\frac{20}{24 \overline{)480}}$$ $$\frac{6}{84 \overline{)504}}$$ $$\frac{6}{59 \overline{)354}}$$

Exercise D9

$$\frac{1}{942 \overline{)942}}$$ $$\frac{2}{491 \overline{)982}}$$ $$\frac{10}{585 \overline{)5,850}}$$ $$\frac{28}{301 \overline{)8,428}}$$ $$\frac{13}{421 \overline{)5,473}}$$

$$\frac{14}{666 \overline{)9,324}}$$ $$\frac{5}{732 \overline{)3,660}}$$ $$\frac{13}{544 \overline{)7,072}}$$ $$\frac{13}{390 \overline{)5,070}}$$ $$\frac{9}{759 \overline{)6,831}}$$

$$\frac{33}{153 \overline{)5,049}}$$ $$\frac{19}{437 \overline{)8,303}}$$ $$\frac{15}{552 \overline{)8,280}}$$ $$\frac{1}{821 \overline{)821}}$$ $$\frac{22}{302 \overline{)6,644}}$$

$$\frac{4}{369 \overline{)1,476}}$$ $$\frac{32}{305 \overline{)9,760}}$$ $$\frac{8}{950 \overline{)7,600}}$$ $$\frac{15}{524 \overline{)7,860}}$$ $$\frac{16}{473 \overline{)7,568}}$$

Exercise D10

$$\frac{1 \text{ R}69}{87 \overline{)156}}$$ $$\frac{11 \text{ R}37}{74 \overline{)851}}$$ $$\frac{29 \text{ R}13}{28 \overline{)825}}$$ $$\frac{25 \text{ R}23}{31 \overline{)798}}$$ $$\frac{6 \text{ R}30}{90 \overline{)570}}$$

$$\frac{16 \text{ R}16}{20 \overline{)336}}$$ $$\frac{1 \text{ R}14}{89 \overline{)103}}$$ $$\frac{10 \text{ R}8}{37 \overline{)378}}$$ $$\frac{2 \text{ R}2}{94 \overline{)190}}$$ $$\frac{5 \text{ R}8}{91 \overline{)463}}$$

$$\frac{4 \text{ R}17}{77 \overline{)325}}$$ $$\frac{24 \text{ R}27}{39 \overline{)963}}$$ $$\frac{17 \text{ R}9}{26 \overline{)451}}$$ $$\frac{16 \text{ R}25}{55 \overline{)905}}$$ $$\frac{32 \text{ R}18}{20 \overline{)658}}$$

$$\frac{71 \text{ R}8}{12 \overline{)860}}$$ $$\frac{42 \text{ R}3}{12 \overline{)507}}$$ $$\frac{7 \text{ R}2}{42 \overline{)296}}$$ $$\frac{40 \text{ R}11}{17 \overline{)691}}$$ $$\frac{5 \text{ R}44}{57 \overline{)329}}$$

Exercise D11

$$\begin{array}{r} 3 \\ 697\overline{)2{,}091} \end{array}$$
$$\begin{array}{r} 40 \\ 240\overline{)9{,}600} \end{array}$$
$$\begin{array}{r} 26 \\ 142\overline{)3{,}692} \end{array}$$
$$\begin{array}{r} 9 \\ 677\overline{)6{,}093} \end{array}$$
$$\begin{array}{r} 93 \\ 102\overline{)9{,}486} \end{array}$$

$$\begin{array}{r} 411 \\ 21\overline{)8{,}631} \end{array}$$
$$\begin{array}{r} 14 \\ 272\overline{)3{,}808} \end{array}$$
$$\begin{array}{r} 18 \\ 329\overline{)5{,}922} \end{array}$$
$$\begin{array}{r} 6 \\ 464\overline{)2{,}784} \end{array}$$
$$\begin{array}{r} 20 \\ 435\overline{)8{,}700} \end{array}$$

$$\begin{array}{r} 13 \\ 542\overline{)7{,}046} \end{array}$$
$$\begin{array}{r} 23 \\ 239\overline{)5{,}497} \end{array}$$
$$\begin{array}{r} 6 \\ 434\overline{)2{,}604} \end{array}$$
$$\begin{array}{r} 176 \\ 35\overline{)6{,}160} \end{array}$$
$$\begin{array}{r} 10 \\ 960\overline{)9{,}600} \end{array}$$

$$\begin{array}{r} 55 \\ 72\overline{)3{,}960} \end{array}$$
$$\begin{array}{r} 3 \\ 877\overline{)2{,}631} \end{array}$$
$$\begin{array}{r} 5 \\ 978\overline{)4{,}890} \end{array}$$
$$\begin{array}{r} 19 \\ 482\overline{)9{,}158} \end{array}$$
$$\begin{array}{r} 12 \\ 377\overline{)4{,}524} \end{array}$$

Exercise D12

$$\begin{array}{r} 6\,R2 \\ 69\overline{)416} \end{array}$$
$$\begin{array}{r} 13\,R6 \\ 54\overline{)708} \end{array}$$
$$\begin{array}{r} 10\,R85 \\ 89\overline{)975} \end{array}$$
$$\begin{array}{r} 20\,R7 \\ 26\overline{)527} \end{array}$$
$$\begin{array}{r} 6\,R26 \\ 32\overline{)218} \end{array}$$

$$\begin{array}{r} 24\,R24 \\ 27\overline{)672} \end{array}$$
$$\begin{array}{r} 4\,R11 \\ 98\overline{)403} \end{array}$$
$$\begin{array}{r} 4\,R30 \\ 35\overline{)170} \end{array}$$
$$\begin{array}{r} 10\,R40 \\ 82\overline{)860} \end{array}$$
$$\begin{array}{r} 4\,R30 \\ 55\overline{)250} \end{array}$$

$$\begin{array}{r} 23\,R8 \\ 28\overline{)652} \end{array}$$
$$\begin{array}{r} 53\,R4 \\ 18\overline{)958} \end{array}$$
$$\begin{array}{r} 4\,R18 \\ 56\overline{)242} \end{array}$$
$$\begin{array}{r} 8\,R50 \\ 86\overline{)738} \end{array}$$
$$\begin{array}{r} 24\,R21 \\ 35\overline{)861} \end{array}$$

$$\begin{array}{r} 3\,R30 \\ 42\overline{)156} \end{array}$$
$$\begin{array}{r} 5\,R11 \\ 79\overline{)406} \end{array}$$
$$\begin{array}{r} 5\,R42 \\ 47\overline{)277} \end{array}$$
$$\begin{array}{r} 16\,R2 \\ 11\overline{)178} \end{array}$$

Exercise D13 - Find the quotient (with remainders)

$$\begin{array}{r} 11\,R4 \\ 62\overline{)686} \end{array}$$
$$\begin{array}{r} 15\,R12 \\ 31\overline{)477} \end{array}$$
$$\begin{array}{r} 3\,R78 \\ 95\overline{)363} \end{array}$$
$$\begin{array}{r} 21\,R12 \\ 34\overline{)726} \end{array}$$
$$\begin{array}{r} 12\,R44 \\ 65\overline{)824} \end{array}$$

$$\begin{array}{r} 4\,R7 \\ 72\overline{)295} \end{array}$$
$$\begin{array}{r} 12\,R10 \\ 45\overline{)550} \end{array}$$
$$\begin{array}{r} 5\,R28 \\ 86\overline{)458} \end{array}$$
$$\begin{array}{r} 4\,R33 \\ 73\overline{)325} \end{array}$$
$$\begin{array}{r} 13\,R11 \\ 58\overline{)765} \end{array}$$

$$\begin{array}{r} 27\,R19 \\ 20\overline{)559} \end{array}$$
$$\begin{array}{r} 9\,R22 \\ 35\overline{)337} \end{array}$$
$$\begin{array}{r} 14\,R34 \\ 61\overline{)888} \end{array}$$
$$\begin{array}{r} 6\,R19 \\ 72\overline{)451} \end{array}$$
$$\begin{array}{r} 9\,R39 \\ 97\overline{)912} \end{array}$$

$$\begin{array}{r} 8\,R30 \\ 57\overline{)486} \end{array}$$
$$\begin{array}{r} 9\,R30 \\ 37\overline{)363} \end{array}$$
$$\begin{array}{r} 9\,R45 \\ 59\overline{)576} \end{array}$$
$$\begin{array}{r} 14\,R54 \\ 57\overline{)852} \end{array}$$
$$\begin{array}{r} 13\,R69 \\ 71\overline{)992} \end{array}$$

www.ingramcontent.com/pod-product-compliance
Lightning Source LLC
Chambersburg PA
CBHW080822180526
45168CB00006B/2548